Advanced and Intelligent Manufacturing in China

Series Editor

Jie Chen, Tongji University, Shanghai, China

This is a set of high-level and original academic monographs. This series focuses on the two fields of intelligent manufacturing and equipment, control and information technology, covering a range of core technologies such as Internet of Things, 3D printing, robotics, intelligent equipment, and epitomizing the achievements of technological development in China's manufacturing sector.

With Prof. Jie Chen, a member of the Chinese Academy of Engineering and a control engineering expert in China, as the Editorial in Chief, this series is organized and written by more than 30 young experts and scholars from more than 10 universities and institutes. It typically embodies the technological development achievements of China's manufacturing industry. It will promote the research and development and innovation of advanced intelligent manufacturing technologies, and promote the technological transformation and upgrading of the equipment manufacturing industry.

Li Li · Qingyun Yu · Kuo-Yi Lin · Yumin Ma · Fei Qiao

Data-Driven Scheduling of Semiconductor Manufacturing Systems

Chemical Industry Press Co., Ltd.

Springer

Li Li
Tongji University
Shanghai, China

Qingyun Yu
Tongji University
Shanghai, China

Kuo-Yi Lin
Tongji University
Shanghai, China

Yumin Ma
Tongji University
Shanghai, China

Fei Qiao
Tongji University
Shanghai, China

B&R Book Program

ISSN 2731-5983 ISSN 2731-5991 (electronic)
Advanced and Intelligent Manufacturing in China
ISBN 978-981-19-7587-5 ISBN 978-981-19-7588-2 (eBook)
https://doi.org/10.1007/978-981-19-7588-2

Jointly published with Chemical Industry Press, Beijing, China
The print edition is not for sale in China (Mainland). Customers from China (Mainland) please order the print book from: Chemical Industry Press.

This Springer imprint is published by the registered company Springer Nature Singapore Pte Ltd.
The registered company address is: 152 Beach Road, #21-01/04 Gateway East, Singapore 189721, Singapore

Preface

Scheduling problem is ubiquitous in industrial engineering; its essence is to seek the maximization of system objectives through the reasonable allocation of limited resources. There are extensive and diverse objective contradictions between the limitation of resources (including material resources and time resources) and the pursuit of objectives (such as output, efficiency, speed, etc.). Therefore, scheduling problem has always been one of the research hotspots in academic and engineering technology circles.

Although there are many kinds of practical performance of scheduling problem, such as manufacturing system scheduling, traffic and transportation scheduling, personnel time scheduling, project schedule scheduling, manufacturing system scheduling is undoubtedly the most concerning and has been studied for the longest time among many scheduling problems. Manufacturing system scheduling is the central problem of enterprise production activities organization and management, which is an effective way to improve enterprise comprehensive benefits. It is of great significance to improve enterprise production management level, save cost, improve service quality, enhance enterprise competitiveness, accelerate the recovery of investment, and obtain higher economic benefits. With the help of advanced production planning and scheduling methods, enterprises can gain greater output and profit and higher rate of return on investment on the basis of no or less increase in investment.

Since Johnson published the first classic paper on production scheduling in 1954, the scheduling problem of manufacturing system has gone through the development process from simple to complex, such as single machine scheduling, multi-machine scheduling, flow-shop scheduling, job-shop scheduling, flexible manufacturing system (FMS) scheduling, and so on. During this period, the accumulation of a large number of related research work and achievements has further established the pivotal position of manufacturing system scheduling in the field of scheduling research. Many early researches in the scheduling field were developed under the promotion of the manufacturing industry, the practical problems arising from manufacturing continue to present new challenges. According to Lawler et al., with the continuous breakthrough of the four basic assumptions of the classical scheduling problem (i.e., single piece processing mode, certainty, operability, and single objective), the research focus on scheduling has gradually shifted from the classical scheduling problem to the new scheduling problem. The semiconductor manufacturing system scheduling discussed in this book belongs to this new type of scheduling problem.

As for the process scheduling problem of semiconductor manufacturing system, the author paid attention to it as early as more than ten years ago. Based on the existing research foundation of flexible manufacturing system, discrete event dynamic system and Petri net, the author has tracked and understood the problems and current situation in this field and carried out some preliminary research. Until recent years, with the gradual deepening of the scientific research work of the author and research teams, they have received funding from a number of scientific research projects such as the National Key Basic Research Program of China (973 Program), the National Natural Science Foundation of China, enterprise cooperation, etc., and the research on the optimal scheduling of complex manufacturing processes has been further carried out systematically. The team has fully investigated and understood the theories and methods in this field, accumulated some research achievements, and cooperated with famous semiconductor manufacturing enterprises to make a beneficial exploration and attempt in the application and implementation of relevant research achievements. This book is written on the basis of these work and achievements, which combines the relevant research achievements in the recent 20 years and the author's years of accumulated experience and attempts to systematically discuss the intelligent scheduling problem of complex semiconductor manufacturing system from theory to method to application.

The intelligent scheduling problem of semiconductor manufacturing is a subject with NP characteristics and high complexity and challenge. With the continuous in-depth research on this kind of problem in the world and in response to the needs of the rise and rapid development of semiconductor manufacturing industry in China, the relevant research is still developing and improving. This book is only a phased summary of the author's study and research in recent years, which will inevitably have inadequacies; please criticize and correct.

Shanghai, China

Li Li
Qingyun Yu
Kuo-Yi Lin
Yumin Ma
Fei Qiao

Contents

Chapter 1
Scheduling of Semiconductor Manufacturing System

Semiconductor is the core of many industrial complete equipment, widely used in computer, consumer electronics, network communication, automotive, industrial/medical, military/government, and other core fields. With the popularity of the concept of "intelligence", the importance of chip industry is becoming more and more significant. To get rid of the "pain of lack of core", China has vigorously supported the domestic semiconductor industry from both the policy and capital aspects, striving for the realization of independent replacement. This chapter mainly introduces the scheduling of semiconductor manufacturing systems and its development trend, including the scheduling process, scheduling characteristics, scheduling types and scheduling methods, evaluation indexes, and scheduling problems.

1.1 Semiconductor Manufacturing Process

The semiconductor industry, as the most important part of the electronic components industry, is mainly composed of four components: integrated circuits (about 81%), optoelectronic devices (about 10%), discrete devices (about 6%), and sensors (about 3%). Therefore, semiconductor and integrated circuits are usually equivalent. Integrated circuits are divided into four main categories by product type: logic devices (about 27%), memory (about 23%), microprocessors (about 18%), and simulator components (about 13%). Semiconductor is a demand-driven market. Over the past four decades, the growth of the semiconductor industry has been driven by a shift from the traditional PC and related industries to the mobile market (including smartphones and tablets). In the future, it is likely to shift to wearables and VR/AR devices. From 2000 to 2015, the annual growth rate of China's semiconductor market led the world, up to 21.4% (the annual growth rate of the global semiconductor market was

L. Li et al., *Data-Driven Scheduling of Semiconductor Manufacturing Systems*, Advanced and Intelligent Manufacturing in China,
https://doi.org/10.1007/978-981-19-7588-2_1

3.6%, of which the Asia–Pacific region was about 13%, the United States was nearly 5%, and Europe and Japan were all lower).In terms of global market share, China's semiconductor market share has increased from 5 to 50%, becoming the core market of the global semiconductor industry [1–3].

In 2015, the three major fields of integrated circuits showed a growth trend. The design industry saw the fastest growth, with sales of $21.57 billion, up 26.55% year by year. Sales revenue of the chip manufacturing industry was $14.67 billion, up 26.54% year by year; Package and test sales were $22.52 billion, up 10.19% from the previous year. In terms of the proportion of industrial chain, the proportion of China's design industry is growing fastest, the proportion of packaging and testing has declined, and the proportion of manufacturing has remained stable. Benefiting from the policy support and the development of domestic economy, the three major IC structures tend to be optimized gradually: in 2015, the proportion of China's IC design accounted for 36.70%, manufacturing accounted for 24.95%, packaging and testing accounted for 38.34%; Chip sales amounted to 90.08 billion yuan, with an increase of 26.5%, 8 percentage points higher than the growth rate in 2014 [4–7].

The IC industry chain can be roughly divided into three major areas: circuit design, chip manufacturing, packaging, and testing. Integrated circuit production process is dominated by circuit design, which requires a variety of high-precision equipment and high-purity materials. Its general process is as follows: design companies provide integrated circuit design schemes, chip manufacturers produce wafers, packaging plants package and test integrated circuits, and sales to electronic product enterprises [8].

The semiconductor manufacturing process can be simply divided into wafer manufacturing and integrated circuit manufacturing. Among them, wafer manufacturing roughly includes several steps of purification of ordinary silica sand (quartz sand), molecular pulling-crystal cylinder (cylindrical crystal), and wafer (cutting the wafer into circular wafers) [9]. Among them, molecular crystallization means that the obtained high purity polycrystalline silicon is melted to form liquid silicon, and the silicon seed of single crystal contacts with the liquid surface and slowly pulls up while rotating. Finally, as the silicon atoms leave the liquid surface to solidify, a column of monocrystalline silicon is formed, which is as pure as 99.999999%. Cutting a wafer refers to cutting a silicon wafer with certain specifications from a single crystal silicon rod. These silicon wafers will be washed, polished, cleaned, inspected by human eyes and machine, and finally be inspected for surface defects and impurities by laser scanning. Qualified wafers will be delivered to chip manufacturers [10].

The IC fabrication process is the focus of this book. It consists of a variety of single processes and consists of three steps: thin film fabrication process, graphic transfer process, and doping process. The specific manufacturing process is shown in Fig. 1.1.

(1) Film preparation technology

Thin film preparation means to grow several layers of films with different materials and thicknesses on the surface of the wafer, and the main processes include

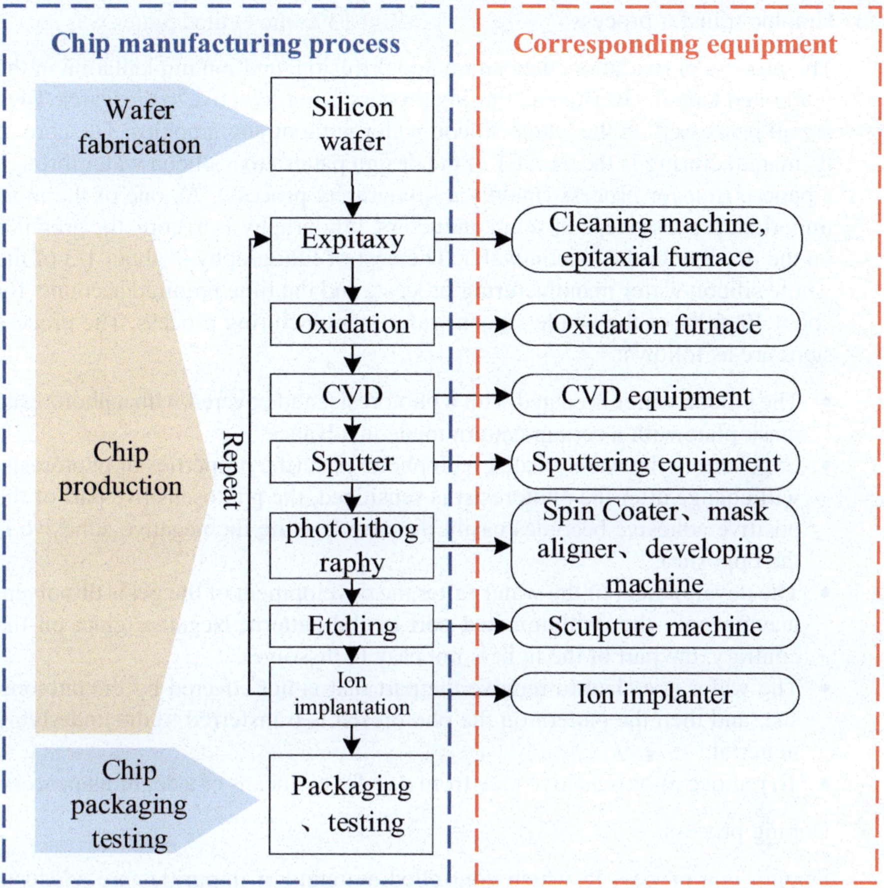

Fig. 1.1 Integrated circuit manufacturing process

three methods: oxidation, chemical vapor deposition (CVD) and physical meteorological deposition (PVD) [11].

- Oxidation: The reaction of a wafer with an oxygen-containing substance (oxygen or an oxidant such as water vapor) at high temperatures to form a thin film of silicon dioxide.
- CVD: one or several compounds or elemental gases containing the elements that constitute the film are passed into the reaction chamber with the substrate, and the solid film is deposited on the surface of the substrate by means of a space vapor phase chemical reaction.
- PVD: The material source is ionized into ions by physical method, and the film with some special function is deposited on the surface of the substrate through the action of low-pressure gas or plasma.

(2) Graphic transfer process

The process of oxidation, deposition and diffusion, and ion implantation in the Integrated Circuit (IC) manufacturing process is not selective to the wafer. They are all processed on the whole silicon wafer without any graphics. The core of IC manufacturing is the transfer of the design pattern to a silicon wafer through a pattern transfer process (mainly a lithography process). As one of the most important process steps of semiconductors, lithography is to copy the graphics on the mask onto the silicon wafer. The cost of lithography is about 1/3 of the whole silicon wafer manufacturing process, and the time required accounts for about 40–60% of the whole silicon wafer manufacturing process. The process steps are as follows:

- The silicon wafer is coated with a photoresist and covered with a photoresist mask plate with a certain pattern made in advance.
- Exposure to the wafer coated with photoresist (the properties of photoresist will change after the photoresist is sensitized, the photosensitive part of the positive adhesive becomes easily dissolved, while the negative adhesive is the opposite).
- The development of the wafer (after the development of the gel is dissolved, leaving only the unilluminated part of the pattern; Negative glue, on the contrary, the part of the light is not easy to dissolve).
- The wafer is etched to remove the part that is not covered by the photoresist, and then the pattern on the photoresist is transferred to the underlying material.
- To remove photosensitive glue from a wafer by means of a degluing process.

(3) Doping process

Doping is a process in which controlled amounts of impurities are added to specific areas of a wafer to alter the electrical properties of a semiconductor. Diffusion and ion implantation are two main processes of semiconductor doping [12].

- Diffusion: The process by which atoms, molecules, or ions are driven by high temperatures (900–1200 °C) from a high concentration zone to a low concentration zone. The concentration of impurity decreases monotonically from surface to body and the distribution of impurity is determined by temperature and diffusion time.
- Ion implantation: In a vacuum system, ions are accelerated by an electric field and changed by a magnetic field so that ions can be injected into the wafer with a certain amount of energy to form an injection layer with special properties in a fixed area and achieve the purpose of doping.

Compared with other manufacturing systems, semiconductor manufacturing systems have the following three distinct characteristics:

(1) Complicated technological process

Production technological process refers to the method and process in which workers use production tools to process all kinds of raw materials and semi-finished products through certain equipment and in a certain sequence, and finally make them into finished products. That is the combination of elements in the production process of products, from raw materials to finished products. The average processing cycle of a silicon wafer in the production line is relatively long, generally about 1 month. The process flow of silicon wafer varies from product to product, from dozens of steps to hundreds of steps, which also leads to the dispersed processing cycle of silicon wafer. A typical process is 250–600 steps, with 60–80 types of equipment used. In addition, there may be different orders and product categories on the production line. There are dozens of products produced on the production line, and there is a large number of re-entrant phenomena in the process, which will make the work-in-process competition for the right to use online equipment very fierce [13].

(2) Multiple entry process

In semiconductor manufacturing, reentrant is the nature of the system. Similar parts at different stages of processing may be processed simultaneously in front of the same equipment, and the parts may repeatedly access some equipment at different stages of the processing process. There are two main reasons for this: first, semiconductor components are hierarchical structures, each layer is produced in the same way, but the added material or accuracy is different; Second, semiconductor processing equipment is very expensive, the need to maximize the use of equipment, resulting in the emergence of multiple into the processing process. In short, the phenomenon of reentrant greatly increases the number of pieces to be processed for each device. In addition, the different types, quantities, and combinations of products, as well as the different complexity of the process flow of each product, make the scheduling and control of the semiconductor production line more complex [14].

(3) Mixed processing mode

Due to the different types of semiconductor production line equipment, its processing methods are also diversified. According to the processing mode of equipment, it is mainly divided into single chip processing, serial batch processing, single card parallel batch processing, and multi-card parallel batch processing. The existence of a hybrid machining mode further increases the complexity of semiconductor production line scheduling. At present, a large number of studies are based on simplified processing methods (single card processing and batch processing).

1.2 Scheduling of Semiconductor Manufacturing System

Production scheduling, as one of the effective ways to improve the economic benefit and market competitiveness of enterprises, is also a research hotspot in industrial engineering, management engineering, automation, and other fields. Generally speaking, production scheduling is to optimize the execution efficiency or cost of a decomposable production task by determining the processing sequence of the workpiece and the allocation of scheduling resources on the premise of satisfying the constraints of technology and resources. As a research proposition with a long history, the requirements of production scheduling include satisfying constraints, optimizing performance and practical efficiency. Its basic tasks can be summarized as modeling and optimization, that is, the understanding of scheduling problems and the solution of scheduling problems. Since the 1950s, scholars at home and abroad have carried out research and exploration on these two tasks, and some advanced scheduling modeling technologies and optimization methods have been put into practice and successfully applied [15].

1.2.1 Scheduling Characteristics

Semiconductor manufacturing system is different from the traditional production mode of job-shop and flow-shop. Whether it is a job shop or flow shop, different processes in the process flow need to be completed on different equipment. A significant feature of semiconductor manufacturing systems is that different processes of the workpiece may repeatedly access the same device, resulting in a large amount of reentry process flow. In the 1990s, Kumar defined semiconductor manufacturing system as the third type of production system developed after the job shop and flow shop—reentrant production system. With the complexity of integrated circuit performance and the miniaturization of component size, semiconductor manufacturing process has become more complex and sophisticated [16]. Therefore, semiconductor manufacturing is recognized as one of the most complex manufacturing systems at present, and its scheduling problem has also been widely concerned by academia and industry and has become a research hotspot in the field of control and industrial engineering.

From the perspective of the scheduling problem, there are two main modes of manufacturing system: Job-shop and Flow-shop. The semiconductor production line differs greatly from these two typical modes because of its obvious reentrant characteristics. The scheduling problem not only has the characteristics of general scheduling problems but also has some obvious special complexity:

(1) Non-zero initial state

As mentioned above, the average processing cycle of semiconductor silicon wafer is longer, generally dozens of days. During this period, new wafers will be put into the production line every day for processing, rather than waiting for all wafers on the production line to be processed and then put in new wafers. Therefore, when the semiconductor production line is initially scheduled, there are already a large number of wafers in the production line, some in the state of being processed, and some in the state of being processed. This non-zero initial state is an important characteristic of scheduling problems in semiconductor manufacturing systems.

(2) On a large scale

The semiconductor production line consists of hundreds of devices, each product process includes hundreds of processes, and there may be as many as hundreds of product types flowing on the production line. This makes the scheduling problem of semiconductor production line much larger than the typical scheduling problem, which makes the complexity of the problem and the difficulty of solving greatly increased. In the production process, the workpieces, equipment, operators, logistics transmission system and buffer zone in the workshop are mutually influenced and restricted [17]. Therefore, not only the loading and processing time of each workpiece, the number of equipment, the buffer capacity of the system, the processing sequence of the workpiece and other resource factors should be considered, but also the uncertain factors such as the operation proficiency of personnel, and even the influence of various dynamic events on scheduling should be considered. Therefore, the scheduling problem of job shop is actually a complex combinatorial optimization problem with many constraints. With the increase of scheduling scale, the amount of computation needed to obtain the feasible solution increases exponentially, and the possibility of obtaining the optimal solution or near-optimal solution becomes less and less.

(3) Uncertainty

The uncertainty of semiconductor production line scheduling is mainly manifested in the following three aspects:

① The total number of tasks is uncertain: in the actual semiconductor production line, new wafers are constantly put into processing every day, but only the number of new wafers invested in a period of time (such as a day or three days) can be known, and the total number of wafers is uncertain.

② Uncertain events: the existence of a large number of uncertain events in the semiconductor manufacturing system usually causes a change in the state of the production line; Therefore, it is necessary to consider the influence of these uncertain events in the scheduling of semiconductor manufacturing systems.

③ Process processing time is uncertain: on the one hand, the same process in different equipment for processing, the required processing time may be different, and with the aging of some parts, the same process on the same equipment processing time will also produce greater changes; On the other hand, some processes need to test pieces (i.e., trial processing) before formal processing, and the test time may change with the change of the number of test pieces, resulting in the uncertain processing time of a card of silicon wafer in this process [18].

(4) Scheduling scheme has short validity

In addition to the current WIP situation on the production line, the formulation of scheduling scheme should also have a detailed and determined releasing plan (determined product type and quantity). Although a certain number of new wafers are put into the production line for processing every day, generally, only a short period of time (such as one day or three days) can be determined for the releasing plan. Therefore, compared with the average production cycle of semiconductor products ranging from ten days to dozens of days, the effective period of the actual semiconductor production line scheduling scheme is generally short. Coupled with the occurrence of a large number of uncertain events, the validity period of the scheduling scheme is difficult to exceed one day.

(5) Local optimization problem

In shop floor production, there are many different production tasks, which may have different requirements for scheduling goals, and sometimes these requirements are mutually exclusive. For example, the requirements of small production cycle, the least overdue orders, the highest equipment utilization rate, and so on. Therefore, how to make the production scheduling system meet these goals as much as possible is also a constant problem for workshop production scheduling [19]. Due to the short validity period of the scheduling scheme, the optimization of the production line system performance index can only be short-term and partial, and only part of the system performance index can be optimized, such as equipment utilization rate, total movement volume, movement rate, etc. However, indicators such as average processing cycle and its variance, just-in-time delivery rate, and delay rate cannot be significantly optimized.

(6) Binding

Constraints are mainly reflected in two aspects: process path constraint and resource constraint. First of all, each product, whether simple or complex, has strict process path constraints, and usually, the sequence of each process cannot be reversed. Secondly, the supply of processing raw materials, the scale of production equipment, the production capacity of production equipment, etc., are not infinite. Therefore, production scheduling is carried out under multiple constraints.

In general, semiconductor manufacturing system scheduling has obvious characteristics of multiple entry, high uncertainty of manufacturing environment, high complexity of manufacturing process and multi-objective optimization of scheduling objectives. Accordingly, dynamic scheduling methods that can respond to real-time operating environment have been paid more attention.

1.2.2 Scheduling Types

- **Scheduling classification of semiconductor production line based on scheduling object**

Semiconductor manufacturing production lines have large scales and different types of equipment, which can be divided into workpiece scheduling, feed control, bottleneck scheduling, batch processing equipment scheduling, production line scheduling, and maintenance scheduling according to different concerns.

(1) Workpiece scheduling

Workpiece scheduling strategy has a direct impact on the performance of the production system, so workpiece scheduling is the focus of semiconductor manufacturing system scheduling research. There are five kinds of scheduling methods: traditional operations research, discrete system simulation, mathematical model, computational intelligence, and artificial intelligence.

(2) Release control

Release control is to determine when and how much raw materials are put into the production system under the guidance of a certain releasing strategy so as to maximize the production capacity of the production system, generally divided into static releasing and dynamic releasing two ways. Static releasing is based on the pre-set rate (such as fixed time interval releasing or random distribution of Poisson flow releasing), because the actual changes of the production line can not be tracked, easy to cause workpiece backlog, so that the performance of the production line decline; Dynamic releasing is based on the actual situation of the production line (such as delivery time and WIP level and other performance indicators), using a heuristic method to control releasing.

(3) Bottleneck equipment scheduling

Bottleneck equipment scheduling refers to the optimal scheduling scheme of the entire semiconductor production line by solving the scheduling problem of the bottleneck area. The loss of capacity at the bottleneck is the loss of the whole plant, so the scheduling problem of bottleneck equipment is very important.

BottLENECK can be identified by observing the queue length of WIP and measuring the utilization rate of machine. ① Analyze the queue length of WIP in the manufacturing system. In this method, either queue length or wait time is measured, and the machine with the longest queue length or wait time is considered to be the manufacturing bottleneck. The advantage of this method is that the instantaneous manufacturing bottleneck of the system can be detected by a simple comparison of queue length or waiting time. The disadvantage is that many production systems have limited WIP queues or none at all, in which case the queue length method cannot be used to detect manufacturing bottlenecks. ② Measuring the utilization rate of different machines in the production system, the machine with the highest utilization rate is the manufacturing bottleneck. However, the utilization rates of different machines are often very similar, so this method cannot say with certainty which machine is the manufacturing bottleneck. Long-time simulations may yield accurate and meaningful results, but this approach is limited by steady-state systems. The method of measuring utilization can not determine the instantaneous manufacturing bottleneck, but only the average bottleneck over a long period of time. It is not appropriate to use it to detect and monitor the transfer of bottleneck.

(4) Batch processing equipment scheduling

Batch processing equipment scheduling is an important part of semiconductor production line scheduling, which has an important effect on the performance of semiconductor production line. Batch processing is the distinguishing feature of semiconductor manufacturing system. The multi-card parallel batch processing equipment, such as oxidized furnace tube, accounts for about 20–30% of the total equipment in the semiconductor manufacturing line, and its scheduling scheme is of great significance in improving the performance of semiconductor manufacturing system.

(5) Production line scheduling

Production line Scheduling pays attention to the flow direction of the workpiece in the whole semiconductor production line and its processing sequence on each equipment, and arranges the processing sequence and start processing time of the workpiece on each processing equipment, namely processing Scheduling, Dispatching, and Sequencing. There are many research methods used in the scheduling of semiconductor manufacturing systems, including traditional operations research, discrete event simulation technology and heuristic rules, as well as advanced artificial intelligence, computational intelligence, and swarm intelligence algorithms.

(6) Equipment maintenance scheduling

Equipment maintenance scheduling is a process used to determine when to remove equipment from the production line for scheduled maintenance. The equipment Maintenance mainly includes Preventive Maintenance (PM) and Corrective Maintenance (CM). The goal of preventive maintenance is to seek a compromise between planned and unplanned downtime. Corrective maintenance refers to the corresponding maintenance of equipment after an unexpected failure. The latter, due to unexpected equipment failure, will lead to higher costs. At present, research on equipment maintenance scheduling mainly focuses on operations research methods or heuristic rules.

- **Scheduling classification of semiconductor production line based on scheduling environment and tasks**

Based on the different scheduling environments and tasks, the scheduling of semiconductor manufacturing systems can be divided into static scheduling and dynamic scheduling.

(1) Static scheduling

Static scheduling refers to the process of forming an optimal scheduling scheme under the premise that the state of the manufacturing system and the processing task are determined. Static scheduling takes place at some point in time t_0. Of the manufacturing system state $U(T_0)$, determined workpiece information (specific processing task description) and time length T_0 (commonly referred to as scheduling depth) is the input, and the appropriate scheduling algorithm is adopted to generate the scheduling cycle under the condition of satisfying the constraint conditions and optimization objectives $[t_0, t_0 + T_0]$ in the scheduling scheme. The constraint conditions of static scheduling include system resources, product process flow, delivery time, etc., and the optimization objective includes evaluation of the processing cycle, delivery time, performance indicators of manufacturing system, such as equipment performance rate, productivity, etc. Once the scheduling scheme is produced, the processing scheme of all the workpieces is determined and will not be changed in the later processing process.

(2) Dynamic scheduling

Dynamic scheduling refers to the process of dynamically generating a scheduling scheme according to the state of manufacturing system and the actual situation of machining task. There are two ways to realize dynamic scheduling: one is to adjust the static scheduling scheme in time and generate a new scheduling scheme based on the existing static scheduling scheme and the field state and processing task information of the manufacturing system. This scheduling process is also called rescheduling; Second, there is no static scheduling scheme in advance, and the processing task is determined for idle equipment directly according to the real-time state of the manufacturing system and the processing task information. This scheduling process is also called real-time scheduling.

Both of the above two methods can obtain a highly operable scheduling scheme, but the optimization calculation process is different. The real-time scheduling usually only considers the local information in the decision-making, so the scheduling scheme obtained is only feasible, which may have a large distance from the optimal scheduling scheme. Rescheduling is based on the existing static scheduling scheme. According to more system state information and processing task information to dynamically adjust the static scheduling scheme, the obtained scheduling scheme is not only operable, but the optimization effect is also better, closer to the optimal scheduling scheme.

Compared with static scheduling, dynamic scheduling can produce a more operable decision-making scheme according to the actual situation of production site. In view of the characteristics of dynamic scheduling, the following two factors must be fully considered: first, the optimization process must make full use of the real-time information reflecting the state of manufacturing system and processing task; second, the dynamic scheduling scheme must be completed within a short time without affecting the operation of equipment.

1.2.3 Scheduling Methods

At present, various scheduling methods can be roughly classified into five categories: the methods based on operations research, the methods based on one-step heuristic rules, and the methods based on artificial intelligence, computational intelligence, and swarm intelligence.

(1) Methods based on operations research

In this method, the production scheduling problem is transformed into a mathematical programming model, and the branch-and-bound method based on an enumeration idea or dynamic programming algorithm is used to solve the optimal or approximate optimal solution of the scheduling problem, which is an accurate algorithm. For complex problems, especially in the semiconductor wafer manufacturing industry with different production characteristics from the traditional Job-shop and Flow-shop, this pure mathematical method has the weaknesses of difficult model extraction, a large amount of computation and difficult algorithm realization. For the dynamic scheduling in the production environment, it is complicated to realize and cannot solve the dynamic and fast response problems.

(2) Methods based on one-step heuristic rules

Heuristic rule refers to the method of choosing some or some attributes of the workpiece as the priority of the workpiece and selecting the workpiece for processing according to the priority level. According to the different scheduling objectives, the semiconductor manufacturing process heuristic rules can be divided into the rules based on the delivery date, the rules based on the processing cycle, the rules based on the waiting time of the workpiece, the rules based on whether the application procedure of the workpiece is the same, and the rules based on the load balancing. Heuristic rules have become the first choice for dynamic scheduling in the actual semiconductor manufacturing environment due to their simplicity and rapidness, but they also have some limitations, which can only improve the individual performance index of the product but have a weak ability to improve the overall performance of the production line.

Due to scheduling optimization of semiconductor manufacturing process is a very complicated problem. Its performance is good or bad depends not only on the scheduling policy itself, and the variance of the system model, processing time, the actual average processing cycle, and processing cycle than the relevant theory, and the system bottleneck in the device number, need to repeat visits, emergency orders to join factors have very close ties. Although heuristic rules require little computation, have high efficiency and good real-time performance, they usually only provide feasible solutions for one or more targets, and lack effective grasp and prediction ability for the overall performance. The scheduling results of heuristic rules may deviate from the global optimization of the system to a certain extent, or even a large deviation. Therefore, heuristic rules often need to be used in conjunction with intelligent methods that select among alternative rules based on the state of the system. Typical research methods usually use a combination of some kind of intelligent method, simulation method, and heuristic rules.

(3) Methods based on artificial intelligence, computational intelligence, and swarm intelligence

Artificial intelligence, also known as machine intelligence, is a comprehensive discipline developed from the interpenetration of computer science, cybernetics, information theory, neurophysiology, psychology, linguistics, and other disciplines. The commonly used artificial intelligence systems in semiconductor scheduling algorithms include expert systems and artificial neural networks, among which the artificial neural network is usually combined with other methods (such as dynamic programming).

Computational intelligence is based on the behavioral patterns of human beings and organisms or the motion patterns of substances. It establishes algorithm models through mathematical abstraction and solves combinatorial optimization problems through computer calculation. The commonly used computational intelligence is tabu search, simulated annealing, genetic algorithm, artificial immune algorithm and so on. In semiconductor manufacturing system scheduling, it is possible to solve scheduling problems by either using a single computational intelligence method, combining different computational intelligence algorithms or combining computational intelligence algorithms with modeling techniques to obtain better performance.

Swarm intelligence is an algorithm and model that is inspired by the swarm behavior of social organisms and simulates the abstraction. Without centralized control and a global model, swarm intelligence provides the basis for finding solutions to complex distributed problems. Common swarm intelligence has an ant colony optimization algorithm, pheromone algorithm, particle swarm optimization algorithm, and so on. The application of swarm intelligence in semiconductor manufacturing system scheduling is relatively rare.

1.2.4 Evaluation Indicators

The purpose of scheduling semiconductor production line is to optimize its system performance. Combined with the characteristics of semiconductor production line, the main indicators used to measure the impact of scheduling results on the performance of semiconductor production line system include:

(1) Yield

The percentage of the total product that is acceptable, often referred to as the percentage of acceptable cores on a silicon wafer. The yield has a significant impact on the economic benefits of semiconductor production lines. Obviously, the higher the yield, the higher the economic benefits. The production rate is greatly affected by the equipment process, and the main effect of scheduling is to shorten the residence time of the workpiece in the workshop as far as possible, reduce the chance of chip contamination, to ensure a higher yield.

(2) Quantity of Work in Process (WIP)

The total number of uncompleted pieces of work in the production line, i.e., the total number of silicon wafers on the production line or the total number of wafers. Minimizing the WIP number is an optimization objective related to minimizing processing cycles. Try to control the WIP number of semiconductor production line to be equal to the expected value, which is related to the processing capacity of semiconductor production line. Even if the WIP number continues to decline below the WIP expectation, the processing cycle will not be significantly shortened; When the WIP quantity is higher than the desired target value, the more WIP quantity, the longer the processing cycle will be. In addition, the higher the number of WIP, the more capital occupied, which will directly affect the economic benefits of enterprises.

(3) Equipment Utility (Machine Utility)

The ratio of the time spent in the processing state to the time spent on the machine can be measured by the idle cost of the equipment. Equipment utilization is related to the number of WIPs. Generally speaking, the number of WIPs is large and the equipment utilization rate is high. However, when the number of WIPs is saturated, the device utilization will not improve even if the number of WIPs increases again. Obviously, the higher the utilization of equipment, the greater the number of processed workpiece, and the greater the value created.

(4) Average processing cycle and its variance

Processing cycle refers to the time taken by a silicon wafer from entering the semiconductor production line to finishing the processing of all procedures and leaving the production line, also known as the wafer flow time. Average processing cycle refers to the average processing time of a multi-card silicon wafer in the same process, and its variance refers to the mean root square of the processing cycle of each card silicon wafer and its average processing cycle. The mean cycle time and its variance indicate the responsiveness and just-in-time delivery of the system.

(5) Total movement

A silicon wafer that has completed one step of processing is called moving one step. The total Movement is the total number of moves (cards · steps) of all silicon wafers in a unit of time (such as a shift, 12 h). The higher the total movement, the higher the number of processing tasks completed by the production line. Movement is an important index to measure the performance of semiconductor production line. The higher its value is, the higher the processing capacity of semiconductor production line is, and the higher the utilization rate of equipment is.

(6) Movement speed

The average number of moves (steps/card) of a silicon wafer per unit time (e.g., 12 h per shift). The higher the moving rate is, the faster the wafer flows in the production line and the shorter the average processing cycle is.

(7) Productivity

The number of cards or silicon wafers flowing out of the production line per unit time (usually shift or day). Ideally, productivity is equal to the rate of feed. It becomes inversely proportional to processing cycle, namely processing cycle is shorter, and criterion productivity is higher. The productivity of a semiconductor production line determines the cost of the final product, the cycle time, and customer satisfaction. Obviously, the higher the productivity, the higher the value created per unit of time, and the higher the processing efficiency of the production line.

(8) On time delivery rate

Percentage of the number of parts delivered on time (on time or ahead of time) as a percentage of the total number of parts completed.

(9) Tardiness

The number of undelivered parts as a percentage of the total number of finished parts. It is obvious that the just-in-time delivery rate is directly or indirectly related to the rate of finished product, productivity, processing cycle, WIP, and equipment utilization. The just-in-time delivery rate and the out-of-date rate are important indexes to measure the advantages and disadvantages of scheduling schemes. Especially, with the continuous intensiveness of competition in semiconductor manufacturing industry, the improvement of just-in-time delivery rate has become an important strategic and tactical index for semiconductor manufacturers to compete for users and occupy the market and has received more and more attention.

The above indexes, which reflect the performance of semiconductor production line system, cannot be optimized at the same time, and the global optimization effect of scheduling scheme on these indexes can only be a compromise or balance in some sense. This is because there are constraints between these performance indicators. For example, to reduce the average processing cycle of the product, the number of WIPs on the production line should be reduced to reduce the waiting time for the workpiece to be processed. Reducing the number of WIP can reduce the capital occupation and indirectly improve the qualified rate of products. However, if the number of WIP is too low, the equipment utilization, total movement, movement rate and productivity of the system will be reduced, and even the just-in-time delivery rate will be reduced, and the profitability of the enterprise will be decreased. On the other hand, if the number of WIP is too high, although the equipment utilization rate can be improved and the total movement volume can be increased, the movement rate may be reduced, the average processing cycle and miss rate will be increased, the qualified rate of products will be reduced, and the working capital of the enterprise will be occupied a lot, affecting the overall profitability of the enterprise. Therefore, a good scheduling scheme should balance the performance indexes and optimize some important indexes as far as possible according to the specific situation, to make the overall performance of the production line reach the optimal or nearly optimal.

1.3 Scheduling Development Trend of Semiconductor Manufacturing System

With the development of modern industrial technology, the manufacturing process, technology, and equipment have become more and more complex. It is difficult to accurately model the system and optimize the system performance through the traditional modeling method based on mechanism model. For example, for a complex silicon wafer production line, although the advanced scheduling idea is used and the scheduling algorithm is carefully designed and implemented, the simulation results are not accurate enough to guide the actual scheduling and scheduling tasks. With the improvement of enterprise information, manufacturing enterprises have significantly improved the real-time and accuracy of data collection, thus promoting the application of data-based methods in the control, online monitoring and fault diagnosis, scheduling optimization and management decision optimization in the manufacturing process. Especially in the field of semiconductor manufacturing, because its key performance indicators cannot be described by mechanism model and monitored online, the data-based prediction method has been widely used. In contrast, the data-based scheduling methods focus more on the combination of data-driven methods and traditional scheduling modeling optimization methods to solve scheduling problems. In this section, three aspects of data preprocessing, data-based scheduling modeling and data-based scheduling optimization in complex manufacturing systems are reviewed.

1.3.1 Data Preprocessing of Complex Manufacturing System

When the manufacturing system reaches a certain scale and the process flow is relatively complex, the automation system will have problems such as a large amount of data, many production attributes, and some noise data contained in the data source. These problems have a significant impact on the performance of data-based scheduling. Therefore, preprocessing the relevant data in the data source is an important part of data-based scheduling. Complex manufacturing data preprocessing mainly focuses on the following three aspects: complex manufacturing data attribute selection, complex manufacturing data clustering, and complex manufacturing data attribute discretization.

(1) Complex manufacturing data attribute selection

Attribute selection can select the more important attribute from the condition attribute. Excessive redundancy of conditional attributes will lead to a decline in the accuracy of classification or regression, unusable rules generated and more conflicts among rules. Common methods of attribute selection include rough set and computational intelligence. For example, Kusiak proposed a method to obtain rules from sample data by using rough sets for quality problems in semiconductor manufacturing and used feature transformation and data set decomposition techniques to improve the accuracy and efficiency of defect prediction. Attribute reduction of rough sets is a NP difficult problem. Chen et al. reduced the search space by using the concept of feature kernel, and then obtained the reduction of attribute sets by using ant colony algorithm, which improved the efficiency of knowledge reduction. Shiue established a two-stage decision tree such as the adaptive scheduling system, the weights of feature selection algorithm based on neural network and genetic algorithm is used for scheduling attribute selection, use self-organizing mapping (Self-Organizing Maps, SOM) for data clustering, the application of decision tree, neural network and support vector machine (SVM) of the three learning algorithm for implementing learning each cluster parameters optimization, improve the generalization ability of the adaptive scheduling knowledge base, and the effectiveness of the proposed is verified by simulation results.

(2) Complex manufacturing data clustering

Clustering is a technique to classify sample data according to similarity, so that similar samples belong to the same class, while samples with low similarity belong to different classes. For large-scale training samples, clustering smoothing noise data can be used. Noise data will affect the accuracy of learning. For example, when C4.5 processes samples containing noise, it will result in a large spanning tree, which will reduce the prediction accuracy, and pruning processing is required. The commonly used methods in clustering include SOM, Fuzzy-C mean, K mean and neural network.

(3) Discretization of complex manufacturing data attributes

Some algorithms and models can only deal with discrete data, such as decision tree, rough set, etc., so it is necessary to adopt attribute discretization technology to transform continuous attribute values into discrete attribute values. For example, Knooce and Li, when mining the optimal scheduling scheme, divide the attribute values by iso-distance discretization according to the characteristics of attribute-oriented protocol algorithm and decision tree. Rafinejad proposed the attribute discretization method based on the fuzzy K-means algorithm, which made the rules extracted from the optimal scheduling scheme better approximate to the optimal scheduling scheme.

Now some complex manufacturing pretreatment technology mainly focus on the attribute selection and data clustering, and in view of the large-scale manufacturing system data, including noise, sample distribution is complex and there are missing, the input variable number, type, more diversity, relationships between input/output variables is nonlinear, strong coupling and so on the characteristic of data preprocessing technology remains to be further in-depth study.

1.3.2 Data-Based Scheduling Modeling

Data-based scheduling modeling includes: (1) generating a model describing the production scheduling process by mapping the data model in the information system; (2) constructing the data-driven prediction model for the uncertain factors of the manufacturing system to realize the refinement of the production scheduling process model; (3) a data-driven performance index prediction model is constructed, and the performance index prediction model can be used to quickly approximate the performance index of the actual manufacturing system and production scheduling process model using scheduling rules in the scheduling environment.

(1) Data-based Scheduling Description Model

The scheduling description model based on data is mainly embodied in Perti network model and discrete event simulation model. The traditional scheduling modeling method is trivial and rigid. If there is replacement of equipment or the introduction of new technology, the whole model needs to be modified. The method based on data can focus the tedious modeling work on the mapping rules from the data model to the production scheduling process model, and the model changes can be easily obtained by modifying the data in the data model, which has better flexibility and expansibility. Gradisar, for example, the production line equipment layout and process flow of processed products, such as data mapping when assigned a Petri net model for describing the process of production scheduling, in the model into some heuristic scheduling rules and evaluate the scheduling performance indicators, in furniture manufacturing process, for example illustrates the feasibility of the method, its shortage is not considering the some dynamic information of production system, cannot be used in semiconductor manufacturing this type of manufacturing system with non-zero initial state of the scheduling model; Mueller semiconductor production line is put forward to the related data mapping method for object-oriented Petri net simulation model, the model of the basic elements of equipment and processing procedures, product process flow, equipment and auxiliary equipment, taking into account the batch processing working procedure, tools and equipment fault time, such factors as the workpiece rework, deficiency is simplified for production line, also does not take into account the characteristics of the nonzero initial scheduling of semiconductor manufacturing system. Ye et al. put forward the dynamic modeling method, which dynamically constructs the discrete event Simulation model of the production line based on the static and dynamic data of the production line, which can reflect the actual working conditions of the production line. However, its shortcoming lies in that the mapping between data and model is targeted at specific Plant Simulation software, and the universality of the conversion method needs to be further improved.

(2) Data driven uncertainty prediction for complex manufacturing systems

Due to the large scale, complexity, and uncertainty of complex manufacturing system, it will face many uncertain factors in the manufacturing process, such as the uncertainty of model parameters, random events, and product quality. How to use the historical data of manufacturing system operation rationally and predict these uncertain factors by data-driven method, to improve the operation accuracy of the description model of manufacturing process is a very practical work.

Many model parameters in complex manufacturing systems are neither fixed values nor specific distributions, but these parameters have important effects on scheduling performance. For example, the workpiece processing time is an important parameter that needs to be used in many scheduling rules. However, in the past work, either the theoretical processing time in the process file is directly used, or the average value is obtained, or the experience based on manual is estimated, but the effect is not ideal. In addition to these modeling basic parameters, many new scheduling strategies also introduce new decision parameters, such as processing cycle, capacity, etc., which are difficult to be estimated by a definite formula, and have a direct impact on the effect of these scheduling policies. Therefore, how to mine the prediction model of these parameters from the historical data is an important part of data-based scheduling.

(3) Data-driven performance index prediction for complex manufacturing systems

For the manufacturing system with large scale and complex manufacturing process, the running time is too long when the production scheduling process description model is run by computer. A semiconductor manufacturing system, for example, involves hundreds of processing equipment, thousands of silicon wafers, and hundreds of processing processes. It takes several hours to run its description model in a one-day scheduling cycle. More convenient to study the scheduling problem of large scale and complicated manufacturing process manufacturing system can be run through the description of the course of its production scheduling model of historical data structure prediction model to predict the performance index of the data driven (e.g., production cycle, wip, yield, etc.), research performance indicators and performance factors affecting environment (scheduling and scheduling strategy).

1.3.3 Data-Based Scheduling Optimization

Data-based scheduling optimization method refers to mining the knowledge that can be used to assist scheduling decision from the optimized scheduling scheme by data mining technology, and its implementation method is consistent with the construction of data-driven prediction model. According to the different ways of generating optimal scheduling schemes, the research of data-based scheduling optimization mainly includes: mining real-time scheduling rules based on schemes obtained by simulation; Real-time scheduling rules are mined by dispatching scheme based on optimization algorithm. Real-time scheduling rules based on offline data mining in information systems.

(1) Scheduling Knowledge Mining Based on Offline Simulation

Many studies have shown that there is no optimal real-time scheduling rule suitable for various types of manufacturing systems. The validity of real-time scheduling rules is directly related to the running state of production line, so the selection of scheduling rules should be guided according to the production scheduling environment. Simulation is one of the important techniques for comparing and selecting scheduling decisions in complex manufacturing systems. In general, there are two simulation ways to select scheduling decisions. One is the Offline simulation, which adopts different scheduling decisions for different production line states, and preserves the scheduling decisions that can best meet the performance indicators, thus constructing the knowledge base. Obviously, the efficiency of these methods is not high, and the generalization ability of the constructed knowledge base is also weak. The other method is online simulation, which adopts different scheduling decisions at decision points to simulate, and selects the scheduling decisions with the best performance indexes to guide real-time dispatching. Online simulation has a strict requirement on the simulation time, which cannot meet the demand of real-time dispatching if not satisfied.

Machine learning can well generalize optimal scheduling decisions and plays a key role in the construction of adaptive scheduling system knowledge base. However, whether offline learning or online learning depends on the scheduling process model of manufacturing system, and the quality of modeling directly affects the learning effect. In addition, the knowledge base acquired by offline learning will degenerate over time, which requires a reasonable updating mechanism. Although the online learning strategy has high robustness, the initial optimization effect is not obvious and the learning speed is slow. How to combine offline learning with online learning to improve the construction of scheduling rule set is a problem worthy of further consideration.

(2) Scheduling Knowledge Mining Based on Offline Optimization

With the development of computer computing power, it is possible to solve large-scale scheduling problems. A bigger bottleneck for solving scheduling problems based on optimization algorithm is that the dispatching scheme is difficult to be implemented due to a large number of uncertain disturbance factors in actual complex manufacturing systems. How to dig out scheduling decisions from a large number of optimization schemes, that is to use appropriate real-time scheduling rules to fit the optimization algorithm, so that the scheduling scheme generated by real-time scheduling rules can better approximate the scheduling scheme of the optimization algorithm, so as to further adapt to the demand of real-time dispatching, is a research of great practical value.

(3) Dispatching Knowledge Mining Based on Offline Data of Information System

Offline data in enterprise information system contains scheduling information and can extract real-time scheduling rules from it. For example, Choi et al. took the multi-entry manufacturing system as the research object, took the scheduling environment of the manufacturing system into consideration, and used decision tree to mine the knowledge of real-time scheduling rule selection suitable for the scheduling environment from offline data. Kwak and Yih using decision tree method from the historical data of manufacturing system offline operation excavated in short operation cycle within different scheduling environment, the real-time scheduling rule choice's influence on the performance indicators, and through the simulation to obtain long-term effective real-time scheduling rules, and consider long-term scheduling performance indicators and short-term scheduling performance refers to the choice of method is applied to real-time scheduling rules. Aiming at the scheduling problem of Flowshop, Murata uses decision tree to obtain the scheduling rules from the actual scheduling scheme and improves the scheduling performance. Qing-qiang guo based on expert experience to determine important degree of each condition attribute in the scheduling knowledge, using rough sets from the production scheduling case in the data mining of relationship between condition attributes and decision attributes, to extract effective scheduling rules, and the scheduling rule acquisition method was applied to one refinery production scheduling knowledge acquisition.

At present, the achievements in the field of data-based scheduling optimization methods are still limited to selecting specific rules from the established real-time scheduling rule set or mining a specific rule offline and applying it to the actual dispatching stage, which is not flexible enough to make real-time adjustment during the production line operation. Oriented production systems are also mainly concentrated in small job shop or flow shop, it is necessary to further study.

1.3.4 Analysis of Research Status

To sum up, with the development of data analysis and data mining technology, data-based methods have been widely used in the modeling and optimization of scheduling problems, which can better overcome the shortcomings of traditional scheduling modeling and optimization methods when solving complex scheduling problems in production processes. However, overall, the research of data-based scheduling method is still in the preliminary stage, and the theoretical and application results have the following limitations:

(1) Based on existing data scheduling research is focused on the data analysis method and the specific scheduling combined modeling and optimization methods, such as improved heuristic scheduling rules by using the method of parameter prediction, the method of using data mining structure adaptive scheduling decision model, the lack of overall scheduling problem based on the data of the solution.

(2) In the existing research on data-based scheduling, the preprocessing of scheduling-related data mainly focuses on data clustering and feature selection, and the attention paid to missing value filling, correlation analysis, constant detection and other data preprocessing methods is insufficient. In addition, there is a lack of corresponding improvement for the defects of common algorithms (such as K-means clustering is sensitive to the initial clustering center). The application of data analysis method in MES and SCADA of actual production system is limited.

(3) In many cases, learning sample data need to be generated through a lot of offline simulation and optimization, so it is difficult to obtain learning sample data for large-scale complex manufacturing system, and there are many variables representing the scheduling environment of complex manufacturing system. It is difficult to construct a learning machine with high generalization ability from such high dimensional small sample learning data.

1.4 Summary

This chapter is a brief overview of the semiconductor manufacturing system, so that the semiconductor production line has a good understanding. To facilitate the understanding of the content of the following chapters.

In the first section, the semiconductor manufacturing process is described in detail. As the whole semiconductor manufacturing system is divided into front end and back end, the front end includes wafer manufacturing and classification, the back end includes packaging and testing, and the front end process is more complex than the back end process. Therefore, the oxidation, lithography, etching and injection of semiconductor manufacturing front-end process are emphatically introduced. Then the reentry flow of semiconductor production line is introduced.

In the second section, the characteristics, types, scheduling methods and evaluation indexes of semiconductor production line scheduling are introduced. It is mainly divided into three parts: scheduling characteristics, scheduling types and methods, and evaluation indexes. Firstly, the characteristics of semiconductor production line scheduling problem are introduced. Secondly, semiconductor scheduling is classified according to scheduling object and scheduling environment, and several common scheduling methods are introduced. Finally, the evaluation indexes are discussed.

The third section mainly introduces the development trend of semiconductor manufacturing system scheduling. It mainly includes complex manufacturing data preprocessing, scheduling modeling based on data, scheduling optimization based on data.

References

1. Wang Z, Wu Q-q (2002) Research on semiconductor production line control and scheduling. Comput Integr Manuf Syst 8(8):607–611
2. Cao G, You H, Jiang Z et al (2008) Dynamic hierarchical planning and scheduling method for semiconductor production line based on TOC. Modul Mach Tool Autom Manuf Technol 2008(10)
3. Shi B, Qiao F, Ma Y (2009) Research on dynamic scheduling of semiconductor production line based on fuzzy Petri net reasoning. Mechatronics 15(4):29–32
4. Wang L, Lu X, Zheng Y (2007) Research on dynamic scheduling of semiconductor production line based on multi-agent technology. Comput Eng 33(13):4–6
5. Wu Q-q, Ma Y-m, Li L et al (2015) Data-driven dynamic scheduling method for semiconductor production line. Control Theory Appl 32(9):1233–1239
6. Monch L, Fowler JW, Dauzere-Peres S et al (2011) A survey of problems, solution techniques, and future challenges in scheduling operations. J Sched 14(6):583–599
7. Ma Y, Qiao F, Chen X et al (2015) Dynamic scheduling method for semiconductor production line based on support vector machine. Comput Integr Manuf Syst 21(3):733–739
8. Jia P, Wu Q-q, Li L (2014) Dynamic dispatching method of semiconductor production line driven by performance index. Comput Integr Manuf Syst 20(11)
9. Su G, Wang X (2011) The establishment of improved Petri net model and optimal scheduling for semiconductor manufacturing system. Syst Eng Theory Pract 31(7):1372–1377

10. Zhang H, Jiang Z, Guo C, Liu H (2006) Real-time scheduling simulation platform for wafer fabrication system based on EOPN. J Shanghai Jiaotong Univ (Chin Ed) 40(11):1857–1863
11. Yao S, Jiang Z, Guo C, Hu H (2008) Ant colony algorithm for optimizing lot processing sequence in wafer manufacturing systems. J Shanghai Jiaotong Univ (Chin Ed) 42(10):1655–1659
12. Li X, Zhou B, Lu Z (2009) An event-driven scheduling algorithm for cluster wafer fabrication. J Shanghai Jiaotong Univ 43(6)
13. Li C, Jiang Z, Li Y, Li N, Geng N, Yao S, Jia W (2013) Application of rule-based batch equipment scheduling in semiconductor wafer fabrication system. J Shanghai Jiaotong Univ (Chin Ed) 47(2):230–235
14. Zhou G, Zhang G, Wang R, Jiang P, Zhang Y (2009) A dynamic scheduling method for cell manufacturing task based on real-time production information is proposed. J Xi'an Jiaotong Univ 43(11)
15. Tan W, Fan Y, Zhou MC et al (2010) Data-driven service composition in enterprise SOA solutions: a Petri net approach. IEEE Trans Autom Sci Eng 7(3):686–694
16. Wei J, Han J, Sun G (2001) Optimal scheduling model for semiconductor manufacturing systems. J Syst Simul 13(2):133–135, 138
17. Zhao T (2015) Multi-performance index prediction method for data-driven semiconductor production line. Master's degree Thesis, Beijing University of Chemical Technology, Beijing
18. Liu X (2015) Research on key parameter prediction model for dynamic scheduling of semiconductor manufacturing process. Master Thesis, Beijing University of Chemical Technology, Beijing
19. Qiao F, Xu X, Fang M et al (2007) Research on performance index system of semiconductor wafer production scheduling. J Tongji Univ: Nat Sci 35(4):537–542

Chapter 2
Data-Driven Scheduling Framework of Semiconductor Manufacturing System

In view of the shortcomings of the current research on complex manufacturing system scheduling, this chapter will introduce a data-based scheduling architecture that is different from traditional scheduling, aiming at the actual large-scale multi-entry complex manufacturing process scheduling problem of semiconductor enterprises, and as an overall data-based solution to solve the large-scale complex manufacturing process scheduling problem, and introduce three application examples.

2.1 Design of Data-Driven Scheduling Framework

The scheduling problem of manufacturing system has been put forward since the 1950s, which has attracted much attention from academia because of its great significance, and the production scheduling system has gradually become an important decision support system for manufacturing enterprises. With the development of research, the scheduling problem of manufacturing system can be subdivided into three categories according to time granularity: production planning, production scheduling, and real-time dispatch. According to the scheduling type, it can be divided into releasing plan, workpiece scheduling, and equipment maintenance scheduling. Mathematical programming, Petri nets, simulation models, heuristic rules, and artificial intelligence methods have been widely used to model and optimize these scheduling problems of different levels and types. These models and methods form different scheduling modules in Production Scheduling System (PSS) according to their corresponding scheduling problems and have been successfully applied in some practical scheduling environments. However, compared with the diversity of theoretical research results of scheduling problems, the application cases of successfully solving scheduling problems in actual manufacturing environment are relatively single, and most of them focus on scheduling simulation systems based on mathematical planning and heuristic scheduling rules, and their intelligence level

L. Li et al., *Data-Driven Scheduling of Semiconductor Manufacturing Systems*, Advanced and Intelligent Manufacturing in China,
https://doi.org/10.1007/978-981-19-7588-2_2

is not high. Taking Intel's Advanced Planning Scheduling System (APS) for semiconductor manufacturing as an example, APS cooperates with production planning module, production scheduling module and real-time scheduling module, and integrates modeling and optimization methods such as simulation model, heuristic real-time scheduling rules and integer programming, which are more traditional. Therefore, there is a gap between the theoretical study of scheduling and the practical application of scheduling system. The main reasons are as follows:

(1) For the specific scheduling problem, the traditional modeling optimization method cannot cope with the large-scale and complexity of the manufacturing system, and the applicability of the corresponding scheduling module is limited.
(2) The research on scheduling problems focuses on the modeling and optimization of specific scheduling problems and the development of corresponding scheduling modules, while the research on cooperation and interaction among PSS modules is less. That is, the research on PSS architecture is insufficient.

Architecture [1] is a description (model) of the basic configuration and connection of each part of the system (whether physical or conceptual objects or entities), that is, "a set of structured, multi-level and multi-view models and methods used to describe different aspects and different development stages of the studied system, which embodies the overall description and understanding of the system and provides tools and methodological guidance for the understanding, design, development and construction of the system". Based on this definition, PSS, which integrates multiple scheduling models or adopts multiple scheduling optimization methods, is defined as PSS architecture. With the increasing complexity of manufacturing system, to improve the availability of scheduling methods, the architecture of scheduling system has been paid attention to since 2000. Pandey et al. [2] put forward the conceptual model of collaborative production scheduling, equipment maintenance and quality control. Monfared and Yang [3] put forward an overall scheme of integrating production planning, production scheduling and control, and based on queuing theory model, integrated the scheduling method of real-time scheduling rules and fuzzy predictive control system, and realized the coordination of production system scheduling and control. Wang et al. [4] proposed a scheduling scheme for scheduling optimization of collaborative capacity planning for semiconductor back-end manufacturing process and introduced capacity constraints as constraints of scheduling optimization planning model through capacity planning model. Lalas et al. [5] proposed a hybrid reverse scheduling method for textile production lines. Firstly, the limited capacity value was obtained by capacity planning model, and the selection of real-time scheduling rules under the constraint of limited capacity was optimized by a discrete-time simulation system. Lin et al. [6] designed a three-tier production planning and scheduling system for a thin film transistor liquid crystal display production line according to three-time granularity of month, day and real-time. In recent years, with the further development and complexity of PSS architecture [7], PSS architecture has the following two ways to realize the optimal scheduling of complex manufacturing systems through the cooperation between scheduling modules:

(1) Multi-agent form: Each scheduling module is encapsulated as an Agent, and the coordination among scheduling modules is carried out in the way of Agent negotiation.

Based on blackboard communication mode, Sadeh et al. [8] realized the collaboration between production planning and scheduling in agile manufacturing. Based on the self-defined multi-agent communication mechanism, Nishioka [9] distributed the production planning Agent scheduling of the manufacturing system cooperatively. Gómez-Gasquet et al. [10] decompose production planning into predictive scheduling (including key performance index prediction) and reactive scheduling and implement predictive scheduling and reactive scheduling through Agent negotiation, thus realizing the cooperation between production planning and production scheduling. Tai and Boucher [11] encapsulate the relevant data of each manufacturing unit and the production scheduling/control method as objects and realizes the coordination of scheduling and control of flexible manufacturing system in a distributed way. Multi-agent-based architecture can integrate simulation model and scheduling module and realize intelligent and adaptive cooperation among scheduling modules through cooperation among agents. The disadvantages are that the existing software development tools do not support the multi-agent system enough, and the multi-agent system leads to slow decision-making speed through negotiation simulation. Therefore, the practicability of the system based on multi-agent is weak.

(2) Component form: The scheduling methods of each scheduling module are encapsulated into components, which are reconstructed according to the characteristics of different manufacturing systems to customize manufacturing systems with high robustness.

Li et al. [12] take the simulation model as the core, and according to the different time granularity, propose a three-tier scheduling module architecture (production planning + production scheduling + rescheduling) or a two-tier scheduling architecture (production planning + real-time scheduling). Several scheduling algorithm components are integrated for each scheduling module. Govind et al. [13] packaged and integrated components such as production planning, near real-time scheduling and real-time dispatching in the OPSched system, and realized the automatic operation of Intel semiconductor manufacturing system by selecting suitable components and achieved high resource utilization. Niu et al. [14] encapsulate the planning model, optimization algorithm and simulation model as components, and realize the Web-based service-oriented scheduling system architecture with high flexibility. The existing integrated development environment provides good support for the development method based on component, so it has better practicability based on component form.

With the development of data mining technology since the 1990s, the data-based method has been applied in the field of manufacturing system scheduling. In view of the limitations of traditional modeling optimization methods, it can be effectively improved by introducing data-based methods. However, in general, the existing data-based scheduling method is a partial improvement of the specific scheduling module for solving specific scheduling problems but does not fully support the existing scheduling system architecture.

In this chapter, aiming at the shortcomings and limitations of traditional scheduling methods, the scheduling problems of complex manufacturing systems are fully supported by the data related to scheduling in manufacturing systems. A data-based scheduling architecture of complex manufacturing system scheduling architecture (DSACMS) is designed, and an application example of DSACMS is illustrated by taking an actual complex silicon wafer processing system as an example.

2.2 Data-Based Scheduling Architecture of Complex Manufacturing System

2.2.1 Overview of DSACMS

As shown in Fig. 2.1, DSACMS consists of four parts, namely, data layer, model layer, scheduling method module and data analysis and processing module.

(1) Data layer

The premise of data-based scheduling is to have abundant data sources related to scheduling. One of the data sources is ERP, MES, SCADA, and other information systems in enterprises. Scheduling-related data from data sources constitute

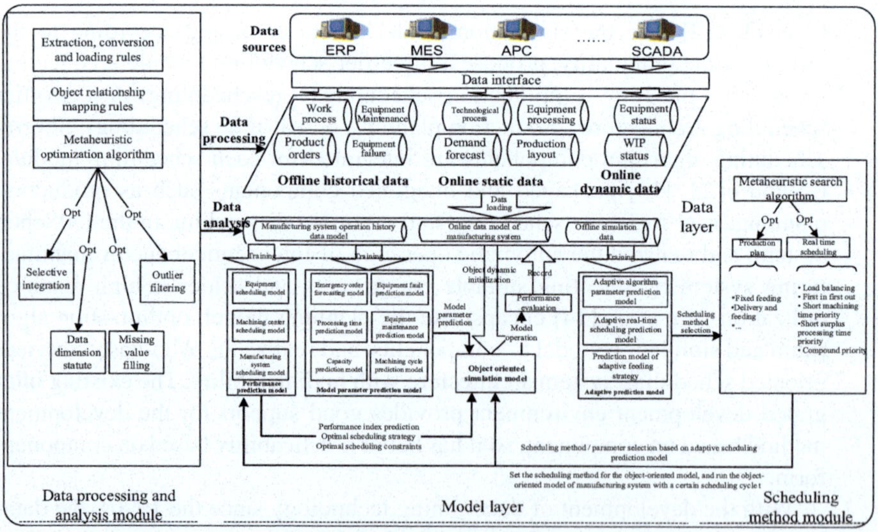

Fig. 2.1 Complex manufacturing system architecture based on data (DSACMS)

the database of the data layer of DSACMS. These data include not only offline historical data (such as workpiece processing history information, product historical production information, equipment historical processing information, equipment maintenance information, equipment failure information, etc.), but also online static data (such as product order information, product process flow information, equipment processing capability information and equipment layout information, etc.) and online dynamic data (such as equipment status information and WIP status information, etc.). The data source can also be offline simulation data generated by offline operation of the simulation model simulating the operation process of the manufacturing system, including offline simulation performance index data and offline simulation optimization scheduling decision data. They are used to construct performance index prediction model, model parameter prediction model and adaptive scheduling model in model layer.

(2) Model layer

The model layer includes an object-oriented simulation model, parameter prediction model, performance index prediction model, and adaptive scheduling model.

① Object-oriented simulation model

Object-oriented simulation model is driven by online data of manufacturing system in data layer through object relation mapping. That is, the object model of simulation model is dynamically constructed according to online data of manufacturing system. The dynamic process of the simulation model, such as the processing mode of the workpiece and the implementation details of the scheduling strategy, are all solidified in the simulation model, while the simulation model object model ensures that the state of objects in the simulation model and the relationship between objects are synchronized with the manufacturing system by dynamically loading the online data of the manufacturing system. To analyze the influence of scheduling decision on scheduling performance index of manufacturing system, scheduling decisions can be evaluated by setting scheduling decisions for object-oriented simulation model and analyzing scheduling performance index of simulation output.

② Data-driven parameter prediction model

Data-driven model parameter prediction model is mainly obtained by mining historical data of manufacturing system operation, such as emergency order, equipment failure, equipment maintenance, processing time, production capacity, processing cycle prediction model, etc. These parameters either represent the probability of occurrence of uncertain events in the model (such as the first four items) or the scheduling parameters of manufacturing system (such as the last two items). Integrating these parameters into the object-oriented simulation model can generate many production system operation sample data considering uncertain information.

③ Data-driven performance prediction model

Data-driven performance prediction model can be obtained by mining offline historical data or offline simulation performance index data. For example, scheduling models of equipment, machining centers and manufacturing systems can be called online and optimized online, which can predict the expected performance indexes and scheduling constraints of equipment, machining centers or manufacturing systems, and provide guidance for real-time selection of optimal scheduling decisions.

④ Data-driven adaptive scheduling model

Data-driven adaptive scheduling model establishes an adaptive scheduling model at the model level through the data of optimal scheduling decision in offline historical data of manufacturing system and the data of optimal scheduling decision in offline simulation. According to the Online scheduling environment of manufacturing system, the adaptive scheduling model is called to complete the actual dispatching operation of manufacturing system. Because the characteristics of the scheduling environment that the scheduling method adapts to are different from the performance indicators concerned. In practical application, the online data (such as equipment status information and WIP status information, etc.) and the performance indicators, scheduling constraints and optimal scheduling decisions obtained by the scheduling model are comprehensively considered, and the appropriate scheduling decisions are selected through the adaptive scheduling model to complete the dispatching.

(3) Scheduling method module

The generation of Offline simulation data depends on the scheduling method module, which includes the production planning module and the real-time dispatching module. Together with the object-oriented simulation model in the model layer and the manufacturing system online data in the data layer, it forms a scheduling method based on simulation. The algorithm component in the production planning module determines the time and quantity of workpieces put into the production line and integrates the releasing rules or algorithms. The rule component in the production scheduling module determines the calculation method of the priority of workpiece processing in equipment, and each scheduling rule optimizes different scheduling performance indicators. The advantage of this method is that it has good real-time performance and can quickly respond to the change of scheduling environment. The disadvantage is that the scheduling environment that releasing strategy and scheduling decision adapt to is different from the performance index concerned, and the optimization degree of manufacturing performance index is over-dependent on the selection of releasing strategy and scheduling decision. Meta-heuristic search algorithm can obtain optimized releasing strategy and real-time scheduling rule configuration by iteratively running simulation models, but it is a time-consuming process to repeatedly run simulation models, especially simulation models of complex manufacturing systems, and it is almost impossible to optimize and select suitable releasing strategy and real-time scheduling rule configuration online through meta-heuristic algorithm. Therefore, in the model layer, a scheme of constructing an adaptive scheduling model by mining the optimal scheduling data of offline simulation is proposed.

The impact of production planning module and real-time scheduling module on performance indicators is related to scheduling period. If scheduling period is short, short-term global performance indicators and short-term local performance indicators are mainly concerned. Performance indicators mainly depend on the selection of initial scheduling environment and real-time scheduling strategy, while production planning and uncertain parameters and events have less impact. If the scheduling period is long, the long-term global performance index is mainly concerned. The performance index mainly depends on the selection of production plan and real-time scheduling strategy, and the influence of uncertain parameters and events must be considered. In contrast, the influence of the initial scheduling environment is weakened. Under different scheduling periods, the influencing factors of performance indicators are different. Corresponding to the model layer, the data-driven performance index prediction model can be divided into real-time performance index, short-term performance index and long-term performance index prediction model according to the simulation model running time of the generated samples. Data-driven adaptive scheduling model can be divided into real-time adaptive scheduling model, short-term adaptive scheduling model and long-term adaptive scheduling model according to the time of running simulation model every time in the process of optimization iteration.

(4) Data processing and analysis module

Data processing and analysis module includes data extraction, transformation and loading rule set to realize data transformation and extraction of scheduling-related attributes. It also includes the mapping rules of data model and object model to realize the mapping between object model and relational model. The core of data processing and analysis module lies in the data preprocessing method and the construction method of data-driven prediction model. Because the data in the information system of manufacturing system is generally noisy, incomplete, highly coupled and irregularly distributed, it is necessary to use data preprocessing technology to filter, purify, denoise and optimize the related Offline data to improve the quality of data mining. Based on the problems of scheduling related data, the data preprocessing module considers outlier filtering, vacancy filling, data dimension reduction and other issues, and iteratively optimizes data preprocessing algorithm parameters such as K-means data clustering, K-means variable clustering and K-nearest neighbor through intelligent optimization algorithm to improve the quality of data preprocessing.

The data-driven forecasting model in the model layer needs to be obtained from the sample data in the data layer by means of data mining. Because scheduling performance index and optimized scheduling scheme need a lot of offline simulation or optimization, to improve generalization ability, computational intelligence method is introduced in generating individual learner and selecting final learner based on selective integration.

2.2.2 Formal Description of DSCAMS

A DSACMS can be defined as a quadruple DSACMS = <Datalevel, Modellevel, Schmodule, DataPrcoAnalyModule>. In which DataLevel represents data layer, ModelLevel represents model layer, DataPrcoAnalyModule is data processing and analysis module, and SchModule is scheduling method module.

(1) Data layer

Definition 2.1 (*Data Model*) The data model R is defined by a set of relational patterns R_1, …, R_{NR}, R = {R_1, …, R_{NR}}, and the relational pattern R_i is defined by attributes $A_{Ri,1}$, …, $A_{Ri,k}$, and R_i = ($A_{Ri,1}$, …, $A_{Ri,NRi}$).

Define K(R_i) = {$A_{Ri,1}$, …, $A_{Ri,NRi}$} as the set of all attributes of R.

Define PK(R_i) ⊆ {$A_{Ri,1}$, …, $A_{Ri,NRi}$} as the primary key to uniquely identify the tuple defined by R.

Define Fk(R_i) {$A_{Ri,1}$, …, A ⊆ $_{Ri,NRi}$} is defined as a foreign key, which is used to associate other relational patterns.

The data layer contains three types of data models, namely manufacturing system online data model R_{MS}, manufacturing system operation history data model R MSRH, learning sample data model R_{LS}, and recording DataLevel = {$R_{MS,}$ R_{MSRH}, R_{LS}}. At scheduling time t, ins$_t$(R_{MS}) is a database instance defined by the online data model R_{MS} of manufacturing system at time t, and the data is extracted from MES and SCADA systems to reflect the current pattern of manufacturing system. ins$_t$(R_{MSRH}) is a database instance defined by the running history data model R MSRH of manufacturing system at current time. Includes the manufacturing system operation record data for a period of time (generally one year or several months) before time t, ins$_t$(R_{LS}) is the database instance defined by the learning sample data model R LS at the current time, In which the data is generated by extracting, transforming and loading (ETL) the database instances ins$_{t'}$(R_{MS}) and ins$_{t'}$(R_{MSRH}) at some time in the past, or running or optimizing the object-oriented simulation model based on ins$_{t'}$(R_{MS}).

① Online data model of manufacturing system

Manufacturing system online data model r ms = {R_{eqp}, R_{wa}, R_{op}, R_{recipe}, R_{proc}, R_{step}, R_{order}, R_{job}}, where:

Equipment is defined by R_{eqp} = (eqp_id, recipe_id, job_id, wa_id, $A_{eqp,1}$, …, $A_{eqp,Ne}$), PK (R_{eqp}) = {eqp_id}, FK (reqp) = {recipe_id, job Wa_id}

$= PK(R_{recipe}) \cup PK(R_{job}) \cup PK \cup (R_{wa})$, eqp_id is the equipment identification, recipe_id indicates the processing menu currently processed by the equipment, job_id indicates the workpiece currently processed, wa_id indicates the processing area where the equipment is located.

The machining area is defined by $R_{wa} = (wa\_id, A_{wa,1}, \ldots, A_{wa,Nw})$, $PK(R_{wa}) = \{wa\_id\}$, wa_id is the machining area identification, $A_{wa,1}, \ldots, A_{wa,NW}$ is the machining area description attribute, describing the machining area name, buffer length and other information.

Operation is defined by r op $= (op\_id, A_{op,1}, \ldots, A_{op,no})$, $PK(R_{op}) = \{op\_id\}$, op_id is operation id, $A_{op,1}, \ldots, A_{op,no}$ is operation description attribute, which describes information such as operation name and operation description.

The equipment processing menu is defined by r recipe $= (recipe\_id, eqp\_id, op\_id, A_{recipe,1}, \ldots, A_{recipe,NR})$, $PK(R_{recipe}) = \{recipe\_id\}$, $FK(R_{recipe}) = \{eqp\_id, Op\_id\} = PK(R_{eqp}) \cup PK(R_{op})$, recipe_id is the processing menu identifier, eqp_id indicates the equipment to which the menu belongs, op_id indicates the process of menu processing, $A_{recipe,1}, \ldots, A_{recipe,NR}$ are the processing menu description attributes, describing the processing time and other information.

Technological process is defined by $R_{proc} = (proc\_id, A_{proc,1}, \ldots, A_{proc,NP})$, $PK(R_{proc}) = \{proc\_id\}$, proc_id is technological process identifier, $A_{proc,1}, \ldots A_{proc,NP}$ is technological process description attribute, describing information such as technological process steps and lithography times.

The work steps of the process are defined by rstep $= (step\_id, proc\_id, oper\_id, position, A_{step,1}, \ldots, A_{step,ns})$, $PK(R_{step}) = \{step\_id\}$. $FK(R_{step}) = PK(R_{proc}) \cup PK(R_{op}) = \{proc\_id, op\_id\}$, where step_id is the process flow to which the process belongs, op_id is the process for processing the process flow, and position is the position of the process flow. $A_{step,1}, \ldots, A_{step,ns}$ is the process description attribute, which describes information such as previous process step, subsequent process step and process constraint.

Order is defined by rorder $= (order\_id, proc\_id, A_{order,1}, \ldots, A_{order,Nor})$, $PK(rorder) = \{order\_id\}$, $FK(A_{order}) = PK(R_{proc}) = \{proc\_id\}$. Order_id indicates the order id, proc_id indicates the technological process required by the order, and $A_{order,1}, \ldots A_{order,Nor}$ are order process description attributes, describing information such as order arrival time, order quantity, delivery date, delivered quantity, expected delivery time, delivered quantity and other order-related external factors.

The workpiece is defined by rjob $= (job\_id, order\_id, eqp\_id, wa\_id, step\_id, A_{job,1}, \ldots, A_{job,NJ})$, $PK(R_{job}) = \{job\_id\}$, $FK(R_{job}) = \{order\_id, eqp\_Wa\_id, step\_id\} = PK(R_{order}) \cup PK(R_{eqp}) \cup PK(R_{wa}) \cup PK(R_{step})$, job_id is the workpiece id, order_id indicates the order to which the workpiece belongs, eqp_id indicates the equipment that is processing the workpiece. Wa_id indicates the processing area where the workpiece is located, step_id indicates the current processing step (when the workpiece is being processed) or the next processing step (when the workpiece is waiting to be processed), $A_{job,1}$, …, $A_{job,NJ}$ are workpiece description attributes, describing information such as the current state of the workpiece.

② Historical data model of manufacturing system operation

The manufacturing system operation history data is described from two angles of equipment operation history information and workpiece operation history information, and defined as $R_{MSRH} = \{R_{erh}, R_{jrh}\}$, where:

Equipment operation history is defined by erh = (eqp_id, event_type, begin_time, end_time, $A_{erh,1}$, …, $A_{erh,Nerh}$), eqp_id is equipment identification, event_type indicates equipment status, such as processing, maintenance, failure, test, etc., and begin_time is the status start. $A_{erh,1}, \ldots, A_{erh,Nerh}$ are state description attributes, such as machining mode, fault type and other information.

Workpiece running history is defined by rjrh = (job_id, event_type, begin_time, end_time, $A_{jrh,1}$, …, A_{jrh}), job_id is the workpiece id, event_type indicates the state of the workpiece, such as processing, waiting, test piece, rework, etc., and begin_time is the state start time $A_{jrh,\,1}$, …, $A_{jrh,Njrh}$ are state description attributes, such as process parameter settings, etc.

③ Learning sample data model

Learning sample data is the basis of constructing a data-driven model. $R_{LS} = \{R_P, R_{UNC}, R_{AS}\}$ is defined. The attributes of relational schema in R_{LS} can be obtained from R_{MS} and R_{MSRH} by using ETL. ETL $\in$ DataPrcoAnalyModule is a set composed of a set of relational algebra operations, which is used to extract attribute values in R_{LS}.

The sample data model of uncertain factors is a set of relational patterns: $\mathbf{R}_{UNC} = \{R_{unc} = (X_{unc,1}, \ldots, X_{unc,\,Nunc}, \mathbf{Y}_{unc}) \mid unc \in UNC\}$, where unc is a collection of uncertain factors in manufacturing system, such as equipment processing time, equipment failure, urgent orders, etc. $\mathbf{X}_{unc} = (X_{unc,1}, \ldots, X_{unc,\,Nunc}) = ETLX_{unc}\ (R_{MS}, R_{ERH}, R_{JRH})$, which is an attribute set (vector) representing the influencing factors of uncertain factors unc, describing information such as equipment continuous running time, equipment switching processing menu frequency, equipment failure, equipment maintenance, etc. $ETLX_{unc} \in$ ETL is a relational algebra for extracting unc influencing factor attributes, $Y_{unc} = ETLY_{unc}\ (R_{MS}, R_{ERH}, R_{JRH})$ is a result attribute (variable) of uncertain factors, describing information such as whether equipment fails or not, etc. $ETLX_{unc} \in$ ETL is a relational algebra for extracting unc result attributes.

Sample data model rp = ($X_{se,1}$, …, $X_{se,Nse}$, $X_{sch,1}$, …, $X_{sch,Nsch}$, Y_{p1}, …, $Y_{p,NP}$), where $X_{se} = (X_{se,1}, \ldots X_{se,Nse}) = ETLX_{se}(R_{MS})$ is an attribute set (vector) representing the current scheduling environment of the manufacturing system, such as work-in-process distribution, urgent workpiece distribution, etc. $ETLX_{se} \in$ ETL represents the relational algebra operation of extracting scheduling environment attributes. $X_{sch} = (X_{sch,1}, \ldots,$

$X_{sch,Nsch}$) indicates the attribute set (vector) of the scheduling method setting scheme, such as scheduling method allocation (by equipment or processing area). Y_{p1} represents the performance index attributes (variables) obtained by the manufacturing system in the scheduling environment X_{se} using the scheduling method to set the scheme X_{sch}, and describes the scheduling performance indexes such as on-time delivery rate and average processing cycle.

The adaptive scheduling sample data model is a set of relational patterns: $\{R_{AS,pi} = (X_{se,1}, \ldots, X_{se,Nse}, Y_{sch,1}, \ldots, Y_{sch,Nsch}) \mid p_i \in P\}$, where $Y_{sch} = (Y_{sch,1}, \ldots, Y_{sch,Nsch})$ indicates the scheduling method setting of optimizing performance index p I under scheduling environment X_{se}.

(2) Model layer

$$\text{ModelLevel} = (\text{OOSM}_{MS}, \text{DDPM}_{MS}).$$

① Object-oriented model of manufacturing system

$OOSM_{MS}$, an object-oriented simulation model of manufacturing system, is an executable simulation model. According to the definition of Object Modeling Technology, OMT), oosmms can be described from three aspects: manufacturing system object model (C_{MS}), manufacturing system dynamic model (D_{MS}) and manufacturing system functional model (F_{MS}).

$OOSM_{MS} = (C_{MS}, D_{MS}, F_{MS})$.

Definition 2.2 (*Object Model*) The object model C is described by a set of class definitions, $C = \{C_1, \ldots, C_{NC}\}$, and the class Ci can be defined in the form of a quadruple, $C_i = \langle A_{Ci}, M_{Ci}, Ref_{Ci}, Agg_{Ci} \rangle$, where:

A_{Ci} is a collection of attributes describing C_i definition objects.

M_{Ci} defines a collection of methods that an object can call for C_i.

Ref_{Ci} refers to the collection of objects defined by C_i, which is denoted as c_i: $C_j \in Ref_{Ci}$, that is, the object C_j defined by c_j (c_j: C_j) is referenced by the object defined by C_i.

Agg_{Ci} contains several sets of objects that make up C_i, which are denoted as c_k s: Set $< C_k > \in Agg_{Ci}$, that is, objects defined by C_i contain a set of objects defined by C_k ($c_k s$: Set$\langle C_k \rangle$).

Id (A_{Ci}) $\subseteq A_{ci}$ is an attribute set that uniquely identifies a C_i object.

Definition 2.3 (*object relation mapping*) ORM $\in$ DataPreprocAnalyModule is defined as the mapping between data model and object model C = ORM (R)

$$ORM(R_i) = C_i$$
$$ORM(K(R_{Ci}) - FK(RCi)) \cup PK(R_{Ci}) = A_{Ci}$$
$$ORM(PK(R_{Ci})) = ID(A_{Ci})$$

$$PK(R_{Cj}) \subseteq FK(R_{Ci}) => ci : ORM(R_C j) \in Ref_{Ci}$$
$$PK(R_{Ci}) \subseteq FK(R_{Ck}) => c_k s : Set < ORM(R_{Ck}) > \in Agg_{Ci}$$

According to the mapping mechanism given in Definition 2.3, the object model C_{MS} of the object-oriented model of manufacturing system can be derived, which describes the types and associations of objects and consists of a group of class definitions. Cms = {C_{eqp}, C_{proc}, C_{op}, C_{wa}, C_{order}, C_{job}, C_{recipe}, C_{step}}, where:

Class C_{eqp} defines equipment objects C_{eqp} = <A_{eqp}, M_{eqp}, Ref_{eqp}, Agg_{eqp}>, where: A_{eqp} = {eqp_id, $A_{eqp,1}$, ..., $A_{eqp,Ne}$} is a set of equipment attributes, eqp_id is equipment identification,$A_{eqp,1}$, ..., Ref_{eqp} = {wa:C_{wa}, recipe:C_{recipe}, job:C_{job}}, wa is the processing area where the equipment is located, recipe is the current processing menu of the equipment, job is the workpiece currently processed by the equipment, Agg_{eqp} = {recipes: Set<C_{recipe}>}, and recipes represent the processing menu set that the equipment can process.

Class C_{wa} defines the machining area object C_{wa} = <A_{wa}, M_{wa}, Ref_{wa}, Agg_{wa}>, where A_{wa} = {wa_id, $A_{wa,1}$, ..., $A_{wa,Nw}$} is the set of equipment attributes, where wa_id is the machining area identification, and $A_{wa,1}$, ..., $A_{wa,Nw}$ is the description attribute; Ref_{wa} = Ø, Agg_{wa} = {eqps: set<C_{eqp}>, jobs: set<C_{job}>}, eqps indicates the equipment set contained in the machining area, and jobs indicates the workpiece set currently located in the machining area.

Class C_{op} defines the operation object C_{op} = <A_{op}, M_{op}, Ref_{op}, Agg_{op}>, where A_{op} = {op_id, $A_{op,1}$, ..., $A_{op,No}$} is the set of equipment attributes, where op_id is the operation id, and $A_{op,1}$, ..., $A_{op,No}$ is the operation description attribute. Ref_{op} = Ø, Agg_{op} = Ø.

Class C_{recipe} defines the processing menu object C_{recipe} = <A_{recipe}, M_{recipe}, Ref_{recipe}, Agg_{recipe}>, where: A_{recipe} = {recipe_id, $A_{recipe,1}$, ..., $A_{recipe,Nr}$} is the set of processing menu attributes, where recipe_id is the processing menu identifier, $A_{recipe,1}$, ..., $A_{recipe,Nr}$ are the description attributes of the processing menu; Ref_{recipe} = {eqp: C_{eqp}, op: Cop}, eqp indicates the equipment where the processing menu is located, op indicates the process of processing menu, Agg_{recipe} = Ø.

C_{proc} class defines the process flow object C_{proc} = <A_{proc}, M_{proc}, Ref_{proc}, Agg_{proc}>, where: A_{proc} = {proc_id, $A_{proc,1}$, ..., $A_{proc,Np}$} is the set of process flow attributes, where proc_id is the process flow id, $A_{proc,1}$, ..., $A_{proc,Np}$ is the process Ref_{proc} = Ø, Agg_{proc} = {steps: Set<C_{step}>} indicates the steps included in the process flow object.

C_{step} class defines a process flow object C_{step} = <A_{step}, M_{step}, Ref_{steP}, Agg_{step}>, where: A_{step} = {step_id, position, $A_{step,1}$, ..., $A_{step,Ns}$} is a set of process attributes. In which step_id is the process identification, position is the position of the process flow to which the process belongs, and $A_{step,1}$, ..., $A_{step,Ns}$ are the description attributes of the process; Ref_{step} = {proc:

Proc}, proc is the technological process to which the process step belongs, $Agg_{step} = \emptyset$.

The C_{order} class defines the process flow object ciorder = <A_{order}, M_{order}, Ref_{order}, Agg_{order}>, where: A_{order} = {order_id, $A_{order,1}$, …, $A_{order,Nor}$} is the set of order attributes, where order_id is the order id, $A_{order,1}$, …, $A_{order,Nor}$ is the description attribute of the order; Ref_{order} = {proc: Proc}, proc is the technological process required to complete the order, Agg_{order} = {jobs: Set<C_{job}>} indicates the artifacts included in the order.

C_{job} class defines the workpiece object C_{job} = <A_{job}, M_{job}, Ref_{job}, Agg_{job}>, where A_{job} = {job_id, $A_{job,1}$, …, $A_{job,Nj}$} is the set of workpiece attributes, where job_id is the workpiece identification, $A_{job,1}$, …, $A_{job,Nj}$ is the description attribute of the workpiece; Ref_{job} = {order: C_{order}, eqp: C_{eqp}, wa: C_{wa}, step: C_{step}}, $Agg_{job} = \emptyset$.

The object instance obj t (C_{MS}) defined by cms can be converted to the database instance ins t(R_{MS}) defined by r ms according to the mapping rule ORM, which is denoted as $obj_t(C_{MS})$ = TRF ($ins_t(R_{MS})$), that is, the online data model instance of manufacturing system at time t, and the object instance and initialization in the object-oriented model can be obtained through the conversion TRF defined by ORM. From the perspective of Model Driven Architecture, MDA), the transformation between model instances can be defined from the mapping between model definitions, so TRF can be defined by ORM. See Fig. 2.2.

The dynamic model of manufacturing system object model, D_{MS}, describes the state transition of objects and the interaction and cooperation between objects, to realize workpiece scheduling and complete orders. There are many ways to describe D_{MS}. For example, Fig. 2.3 describes the scheduling process of semiconductor equipment based on color Petri nets [15]; Fig. 2.4 describes device state switching based on state diagram.

Looking at the functional model F_{MS} from the perspective of scheduling, F_{MS} can be defined as follows:

When the scheduling period t is given, F MS describes the mapping relationship between X_{sch} configured by scheduling method and performance index in scheduling environment X_{se}, that is, $F_{MS} = \{Y_{pi} = f_{pi}(X_{se}, X_{sch}) \mid p_i \in P\}$.

② Data-driven forecasting model

Data-driven forecasting model DDPM in the model layer includes three types of models, namely, Uncertain Factor Estimation Model (UPM),

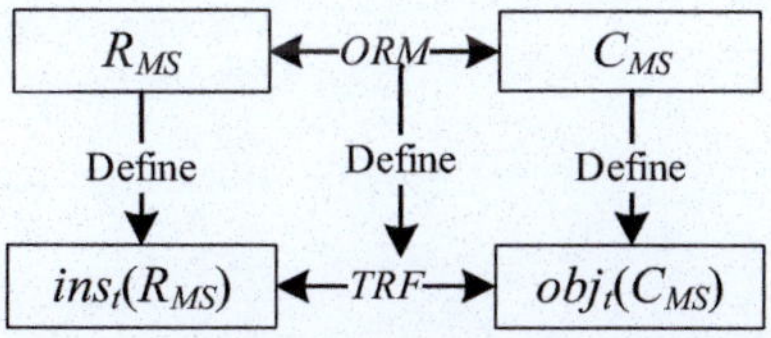

Fig. 2.2 Mapping and transformation of data model and relational model

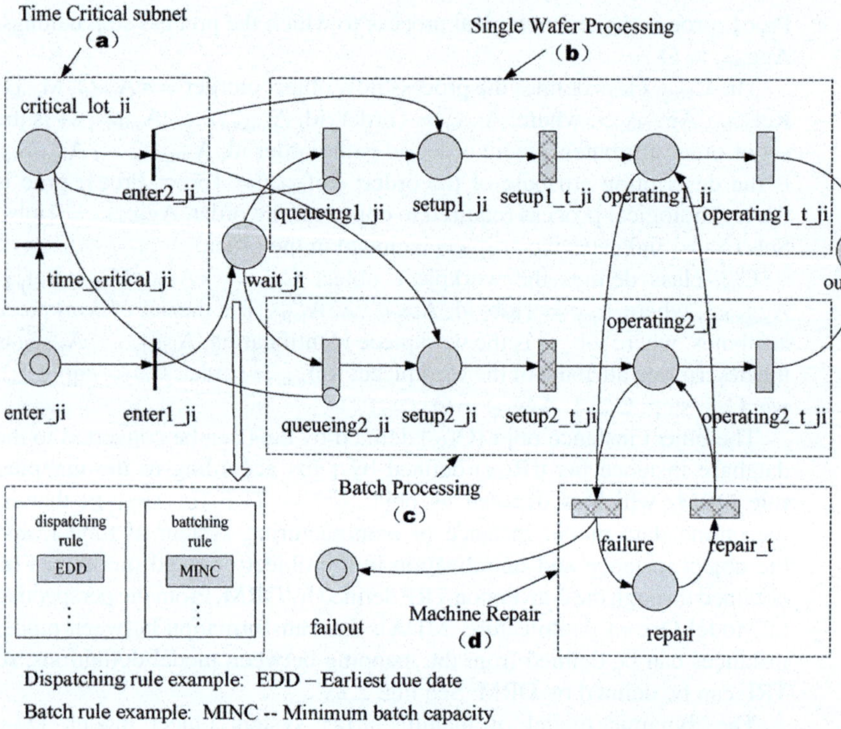

Fig. 2.3 Dynamic model of semiconductor equipment manufacturing process based on Petri Net

Fig. 2.4 Dynamic model of equipment state based on state machine diagram

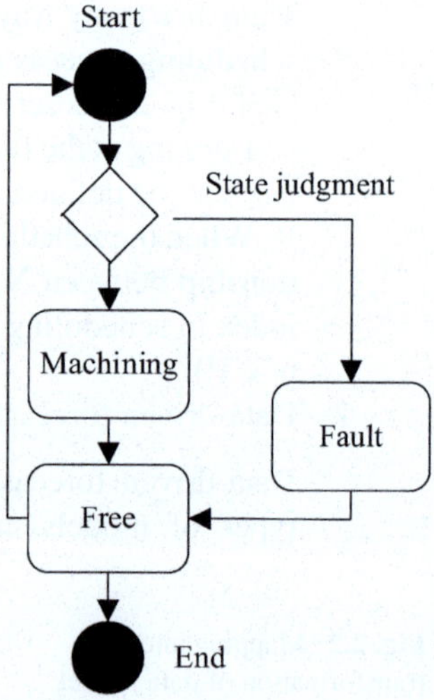

Performance Index Prediction Model (PPM), and Adaptive Scheduling Model (ASPM). That is, DDPM = (PPM, UPM, ASPM). DDPM is constructed by using the sample learning data defined by DataLevel through the method in DataPrcoAnalyModule. PreprocData $\in$ DataPrCoanalY-Module is the data preprocessing method, and BuildRedefinition Model $\in$ DataPrCoanalYModule is the data-based prediction modeling method.

$OOSM_{MS}$ contains uncertain factors (unc), such as workpiece processing time, equipment failure, urgent orders, etc. to make the operation result of $OOSM_{MS}$ more accurate, a set of data-driven uncertain factor estimation models UPM can be obtained from the actual operation history of manufacturing system to characterize these uncertain factors. for unc $\in$ UNC, its data-driven prediction model f′ unc can be learned from $ins_t(unc)$:

$$\begin{aligned} ins_t(Runc) = \{< x_{unc,t'}, y_{unc,t'} > |x_{unc,t'} \\ = ETLX_{unc}(ins_{t'}(R_{MS}), ins_{t'}(R_{ERH}), ins_{t'}(R_{JRH})), \\ y_{unc,t'} = ETLY_{unc}(ins_{t'}(R_{MS}), ins_{t'}(R_{ERH}), ins_{t'}(R_{JRH}))\} \end{aligned}$$

$X_{unc,t'}$ is the value of influencing factor vector x unc of unc, and $Y_{unc,t'}$ is the value of unc result variable y unc. Generally speaking, $y_{unc,t'}$ can be extracted from historical data, and when $y_{unc,t'}$ is difficult to obtain, it can also be obtained by $OOSM_{MS}$ simulation of actual manufacturing system operation.

$Ins_t(R_{unc})$ is preprocessed by preProcData, and the data-driven prediction model f’ unc (X’$_{unc}$) with uncertain factors unc can be obtained by calling the data-driven modeling method BuildPredictionModel:

$$Y_{unc} = f'_{unc}\left(X'_{unc}\right) = BuildPredictionModel(preProcData(ins(R_{unc}))$$

X'_{unc} is the influencing factor of unc after data preprocessing.

Therefore: UPM = $Y_{unc} = f'_{unc}\left(X'_{unc}\right)$ | unc $\in$ UNC}.

When the manufacturing system is large-scale and the manufacturing process is complex, $OOSM_{MS}$ runs for a long time and is difficult to run online. Given the functional model F_{MS} of $OOSM_{MS}$, a set of data-driven performance index prediction models PPM can be obtained from the operation history data of $OOSM_{MS}$ as the approximate expression of F_{MS}, and the approximate output of F_{MS} can be quickly obtained by calling the models in PPM. As to the performance index pi, its data-driven prediction model f′pi is obtained by $ins_t(R_P) = \{<x_{se,t'}, x_{sch}, y_{p1,t'}, \ldots, y_{pNP,t'}>| x_{se,t'} = ETL_{se}(ins_{t'}(R_{MS})), y_{pi,t'} = f_{pi}(x_{se,t'}, x_{sch}), p_i \in P, t' < t\}$.

$x_{se,t'}$ is the value of scheduling environment vector Xs_e at t′, which is extracted from database instance $ins_{t'}(R_{MS})$, and X_{sch} is the value of scheduling method setting vector X_{sch}, which can be specified by the user or traversed by enumeration. $Y_{pi,t'}$ is the value of performance index $p_{i,t'}$ obtained by running oosms in a given scheduling period t when the scheduling environment is $x_{sch,t'}$. The value of pi can also be obtained directly from the database instance of the manufacturing system at t times after t′, that is, the value of pi and t' can be obtained from $ins_{(t'+T)}(R_{MSRH})$, which is denoted as $p_{i,t'} = ETLY_{pi}$ $(ins_{(t'+T)}(R_{MSRH})$.

The $ins_t(R_P)$ is preprocessed by preProcData, and the data-driven prediction model $f'_{pi}(x'_{se}, x_{sch})$ of performance index p_i can be obtained by calling the data-driven modeling method BuildPredictionModel:

$$Y_{pi} = f'_{pi}(X_{se'}, X_{sch}) = \text{BuildPredictionModel}(\text{preProcData}(ins_t(R_P))$$

Xse′ is the scheduling environment vector after data preprocessing protocol.

Therefore: $PPM = \{Y_{pi} = f'_{pi}(X_{se'}, X_{sch}) | p_i \in P'\}$.

In which: P′ is the scheduling performance index set after data preprocessing protocol.

Due to the complexity of manufacturing system, the scheduling method setting of iterative optimizations $OOSM_{MS}$ cannot be completed online. Given the functional model F_{MS} of $OOSM_{MS}$, the data-driven adaptive scheduling model can be obtained from the optimized operation history data of $OOSM_{MS}$ as a method to optimize F_{MS}. As to the performance index p_i, its data-driven adaptive scheduling model argmin $X_{sch'} = f_{pi}(X'_{se}, X_{sch})$ is learned from $ins_t(R_{AS},p_i)$:

$$ins_t(R_{AS,pi}) = \left\{ < x_{se,t'}, y_{sch,t'} > | \; y_{sch,t'} = \arg\min_{Xsch} f_{pi}(X'_{se}, X_{sch}) \right\}$$

Xse,t′ is the value of scheduling environment vector X_{se} at t′, which is extracted from database instance $ins_{t'}$ (R_{MS}). $Y_{sch,t'}$ is a scheduling method setting for optimizing (minimizing optimization) the scheduling performance index p_i by iteratively running $OOSM_{MS}$ under the scheduling environment xse,t′. Ysch,t′ can also be set as a scheduling method for obtaining a better scheduling performance index p_i on the actual production line under the scheduling environment $x_{se,t'}$, that is, when the performance index p i of the manufacturing system reaches a better result (inferred from $ins_{(t'+T)}(R_{MSRH})$ in the time period from t′ to t′ + t), the scheduling method adopted by the manufacturing system at t′ is set as $y_{sch,t'}$.

The $ins_t(R_{AS,pi})$ is preprocessed by preProcData, and the data-driven prediction model argmin Xsch fpi(Xse′, Xsch) with adaptive optimization performance index pI can be obtained by calling the data-driven modeling method BuildPredictionModel:

$Y_{sch} = \text{argmin}x'_{sch} f_{pi}(X_{se'}, X_{sch}) =$ BuildPredictionModel (preProcData(ins(R_{AS}))), where x'_{se} is a scheduling environment variable after data preprocessing protocol.

Therefore: $ASPM = \{Y_{sch} = \text{argmin}_{Xsch} f_{pi}(X'_{se}, X_{sch}) \,|\, p_i \in P\}$.

(3) Scheduling method module

Scheduling method module includes three types of methods, namely, production planning method set, production scheduling method set, and meta heuristic search method set (MHS)

$$\text{SchModule} = \{\text{PlanMethods}, \text{SchMethods}, \text{MHS}\}.$$

The method in PlanMethods is a production planning method for processing orders, such as the releasing strategy in semiconductor manufacturing system, which can adopt fixed releasing, releasing based on due date, multi-objective releasing, intelligent releasing and other methods, and SchMethods realizes the production scheduling method of workpiece scheduling, such as real-time scheduling rules for calculating workpiece priority or search method for workpiece sequencing in semiconductor manufacturing system. The methods in PlanMethods and SchMethods can be implemented in ooms: the methods in PlanMethods can be implemented as member methods of corrder class (implemented in miorder), and the methods in SchMethods can be implemented as member methods of C_{eqp} class, C_{wa} class, or C_{job} class (implemented in M_{eq}p, M_{wa} or M_{job}). The methods in PlanMethods and SchMethods can also be packaged into components for oosms to call. $OOSM_{MS}$ encodes scheduling method settings (for example, allocating real-time scheduling rules according to equipment/production lines in processing areas) in a certain form, which is represented by vector X_{sch}. The value of X_{sch} can be given by enumeration traversal or user setting. $OOSM_{MS}$ decodes X_{sch} according to coding rules and calls corresponding production planning methods to realize scheduling, complete processing, and obtain corresponding performance indicators. When learning samples of data-driven adaptive scheduling model are generated by iteratively running ooms, when the dimension of X_{sch} is high, $\text{argmin}_{Xsch} f_{pi}(Xs_{e,t'}, X_{sch})$ is difficult to realize. Therefore, a meta heuristic search method can be provided by MHS, and a better x_{sch} value can be obtained as a training sample by iteratively optimizing the operation of ooms, that is, mhs ∈ MHS, so that:

$$\text{mhs}_{Xsch}\left(f_{pi}\left(x_{se,t'}, X_{sch}\right)\right) \approx \text{argmin}_{Xsch} f_{pi}\left(x_{se,t'}, X_{sch}\right)$$

(4) Data processing and analysis module

Scheduling method module includes five types of methods, namely, ETL, ORM, Preprocess Data, Built Modeling Model, and MHO.

$$\text{DataProAnalyModule} = \{\text{ETL}, \text{ORM}, \text{PreProcData}, \text{BuildPredictionModel}, \text{MHO}\}$$

The method in ETL is used for data model transformation, ORM realizes the mapping between online data model of manufacturing system and object model in object-oriented model, PreProcData realizes the preprocessing of learning sample data, and BuildRedefinition Model learns from learning sample to obtain data-driven prediction model. MHO optimizes parameters for the shortcomings of parameter sensitivity of methods in PreProcData and BuildRedefinition Model. For example, if it is known that data set DS, preProcData pars ∈ Preprocdata, Pars is the parameter set required by Preprocdata, mho ∈ MHO can be selected to optimize Pars setting through mho (preProcData pars (DS)) to improve the data preprocessing quality of preProcData pars.

2.2.3 DSACMS-Based Modeling and Optimization of Scheduling for Complex Manufacturing Systems

DSACMS supports scheduling modeling and optimization of complex manufacturing systems through the model in ModelLevel and the method in SchModule, as shown in Fig. 2.5. Based on the characteristics of data, it is embodied by DDPM, and UPM, PPM, and $OOSM_{MS}$ jointly support scheduling modeling, in which UPM refines $OOSM_{MS}$ and PPM approximates F_{MS} ASPM and SchModule realize scheduling optimization together, and ASPM realizes adaptive setting of methods in SchModule to improve the intelligent level of scheduling.

Data in the data layer can initialize the object model defined by C_{MS} through TRF defined by ORM mapping and can be used as the medium of model conversion to realize model conversion between F_{MS} and PPM and ASPM. The quality of model transformation depends on the method in DataProcAnalyModel to learn the model with high generalization ability from the data.

Fig. 2.5 DSACMS support for scheduling modeling and optimization

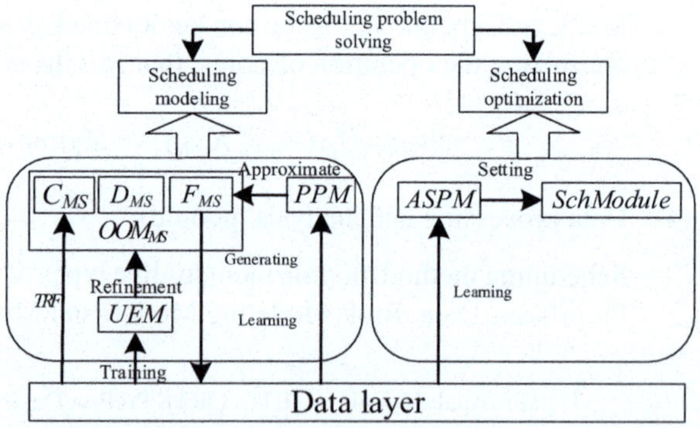

2.2.4 Key Technologies in DSACMS

(1) $OOSM_{MS}$ modeling technology

According to the definition of the functional model F_{MS} of $OOSM_{MS}$, the purpose of running $OOSM_{MS}$ is to obtain the influence of different scheduling methods on scheduling performance indicators in a specific scheduling environment. Therefore, $OOSM_{MS}$ is of great significance to DSACMS. $OOSM_{MS}$ object model CMS and $OOSM_{MS}$ function model FMS have been defined in the formal description of DSACMS. The mapping relationship between data model and object model ORM can initialize the object model defined by C_{MS} according to the current data in MES and SCADA system, so that $OOSM_{MS}$ can reflect the actual working conditions of manufacturing system. Therefore, the modeling of $OOSM_{MS}$ also needs to complete the following two tasks: (1) Define the attributes of classes in C_{MS} according to the actual situation of manufacturing system; According to the processing flow of manufacturing system, the dynamic model D_{MS} of $OOSM_{MS}$ is designed. The high-quality $OOSM_{MS}$ model can make F_{MS} approximate to the operation results of the actual manufacturing system.

(2) The design of schmodule

SchModule contains production planning and scheduling methods applied to $OOSM_{MS}$. If the methods in SchModule are applicable to $OOSM_{MS}$, the efficiency of collecting learning samples by running $OOSM_{MS}$ can be greatly improved. Better results can also be obtained by adaptively selecting the scheduling scheme settings based on the data-driven forecasting model.

(3) Design of DataProAnalymodule

$OOSM_{MS}$ and SchModule play an important role in DSACMS, but they still belong to the category of traditional scheduling modeling optimization. The key of DSACMS is to construct a prediction model with high generalization ability and support the scheduling of complex manufacturing systems in uncertain factors prediction, performance index prediction and adaptive scheduling prediction. The design of DataProAnalyModule plays a key role in DSACMS. ORM has been defined by DSACMS, and ETL can be implemented by using Structured Query Language, SQL) in the remaining chapters of this paper. The meta-heuristic optimization algorithm (MHO) based on computational intelligence will be used to optimize the parameters of data preprocessing method (preProcData) and data modeling method (BuildRedefinition Model) and be used for data preprocessing and data-driven prediction model construction of complex manufacturing systems.

2.3 Application Examples

2.3.1 *Overview of Fabsys*

Semiconductor manufacturing system is a production line with monocrystalline silicon as raw material and integrated circuit as product. The manufacturing process is shown in Fig. 2.5. After slicing and polishing monocrystalline silicon ingot (Fig. 2.5 and 1.2), the silicon wafer is processed by front-end process (Figs. 2.5 and 3.8) and back-end process (Figs. 2.5 and 2.9) to make integrated circuit chip. The front-end technology is silicon wafer processing technology, including oxidation, lithography, etching, ion implantation, diffusion, cleaning, and other processes. The back-end process divides, packages, and tests the silicon wafer. Compared with the back-end process, the front-end process has many steps, complicated process flow and high equipment cost. The scheduling problem of the silicon wafer processing production line dealing with the front-end process is the research object of this paper. The scale of silicon wafer processing production line can reach hundreds of equipments, and each product needs to complete hundreds of processing procedures. Because the final integrated circuit needs to form several circuit layers on the silicon wafer, the silicon wafer will repeatedly visit the same equipment during the silicon wafer processing, that is, reentrant phenomenon will occur, so the semiconductor manufacturing system is called reentrant production system.

Shanghai Silicon Wafer Manufacturing Co., Ltd. is a leading analog chip foundry in China, which is a high-tech enterprise engaged in IC design, manufacturing, sales, and technical services. The basic parameters of the 5-inch and 6-inch silicon wafer processing lines (FabSys) of this enterprise are shown in Table 2.1. According to the data in Table 2.1, FabSys has the characteristics of complex process flow, multiple inputs, multi-product mixed processing, and various types of equipment processing. In addition, uncertain factors such as equipment failure, order change, and rework frequently occur during the processing of FabSys. Therefore, FabSys is a typical complex manufacturing system.

The processing procedure of equipment processing in FabSys corresponds to the front-end processing technology 3–8 in Fig. 2.6. These equipments are divided into 8 processing areas according to their functions. The names and abbreviations of these

Table 2.1 Basic parameters of Fabsys production line

Basic parameters (unit)	Orders of magnitude
Equipment scale (units)	$\geq$ 500
Equipment processing type (class)	5
Product type (class)	$\geq$ 100
Processing flow steps (steps)	$\geq$ 300
Lithography times (times)	$\geq$ 5
Scale of work in process (piece)	$\geq$ 40,000

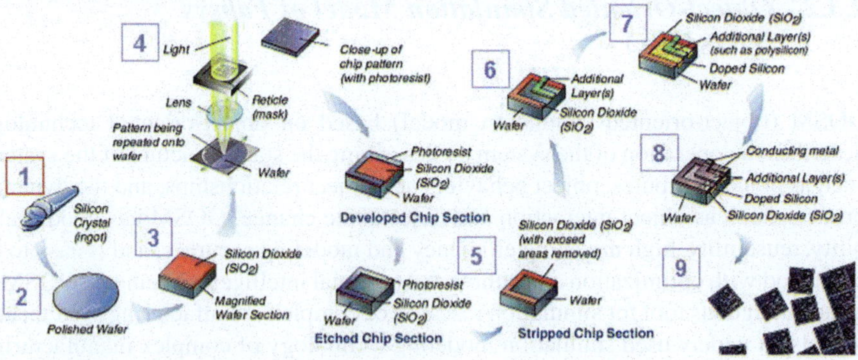

Fig. 2.6 Semiconductor manufacturing processes

Table 2.2 Processing zones and abbreviations in Fabsys

Name of processing zone	Abbreviation of processing area
Oxidation diffusion region	DF
Injection area	IM
Epitaxial region	EP
Lithography area	LT
Dry etching area	PE
Deposition area	PC
Sputtering area	TF
Wet etching, wet glue removal and wet cleaning	WT

processing areas are shown in Table 2.2, and the collection of all processing areas is denoted as Work_Areas = {DF, IM, EP, LT, PE, PC, TF, WT}.

In this paper, the scheduling problem of 5-inch and 6-inch silicon wafer processing lines (FabSys) mentioned above is taken as the verification object. From the equipment and WIP scale of FabSys in Table 2.1, the scheduling problem of FabSys is a large-scale and non-zero initial state scheduling problem. In the scheduling process, it is difficult to obtain a long-term effective scheduling scheme that optimizes the global performance index due to uncertain events such as urgent orders and equipment failures and uncertain parameters such as processing time and remaining processing cycles. In this paper, the scheduling simulation model of silicon wafer processing production line ($OOSM_{fab}$), which was independently developed by the research group in the early stage, can load the actual online production data of FabSys in real time through the data interface, and can simulate the running status of enterprise production lines. $OOSM_{fab}$ implements heuristic scheduling rules and enterprise general scheduling rules and determines the priority of silicon wafers on equipment through these scheduling rules, thus generating scheduling schemes.

2.3.2 *Object-Oriented Simulation Model of Fabsys ($OOSM_{fab}$)*

OOSM (object-oriented simulation model) based on object-oriented technology describes the operation of the system by describing the static structure of the system, such as object attributes, object behaviors and object relationships, and the dynamic process, such as object interaction and object state change. OOSM has good scalability, reusability, high modeling efficiency and modeling accuracy, and is easy to be combined with optimization algorithms and artificial intelligence methods. OOSM is an advantageous tool for simulation research of complex manufacturing system, and it is also a widely used simulation modeling technology of complex manufacturing system. Unified modeling language (UML) is the most widely used object-oriented analysis and modeling language. In this section, based on full investigation of FabSys and ignoring some enterprise implementation details, the object-oriented modeling of FabSys is carried out based on UML, and the composition structure and operation process of FabSys are described from three aspects: object model, dynamic model, and functional model. $OOSM_{fab}$, the oosm system of FabSys, is realized based on the discrete modeling and simulation tool Plant Simulation and its object-oriented programming language Sim talk. The process described by the dynamic model of FabSys is encapsulated as the Method in Sim talk, which is implemented and solidified in $OOSM_{fab}$, and TRF is loaded into the online data model through ORM mapping, keeping $OOSM_{fab}$ and FabSys synchronized.

(1) Object model of $OOSM_{fab}$ (C_{fab})

With reference to C_{MS}, C_{fab} can be defined by class diagram. In the Process of FabSys processing, the core classes of modeling are process, Equipment and Lot. The FabSys class diagram constructed by these three core classes is shown in Fig. 2.7.

The Process class defines the silicon wafer Processing process, the process_ID is the process number, each process object contains several processing Steps, and the step class defines the processing steps, each of which is determined by the processing operation and its relative position in the process flow. Order class defines the customer order, including the silicon wafers ordered by the customer and having the same Process, and defines the required quantity and due_date of silicon wafers. The Release_Plan class defines the releasing plan of an order, including its corresponding order, release_time and quantity.

The Equipment class defines silicon wafer processing equipment, each equipment is assigned a unique equipment number (eqp_ID), and each equipment has several processing menus to handle different processing procedures. Recipe class defines the processing menu, and eqp_ID indicates the equipment number to which the processing menu belongs. In addition, the processing menu also includes two attributes: operation and processing_time. Therefore, the processing time is not unique when different equipment processes different processing procedures, which reflects the interchangeability and different

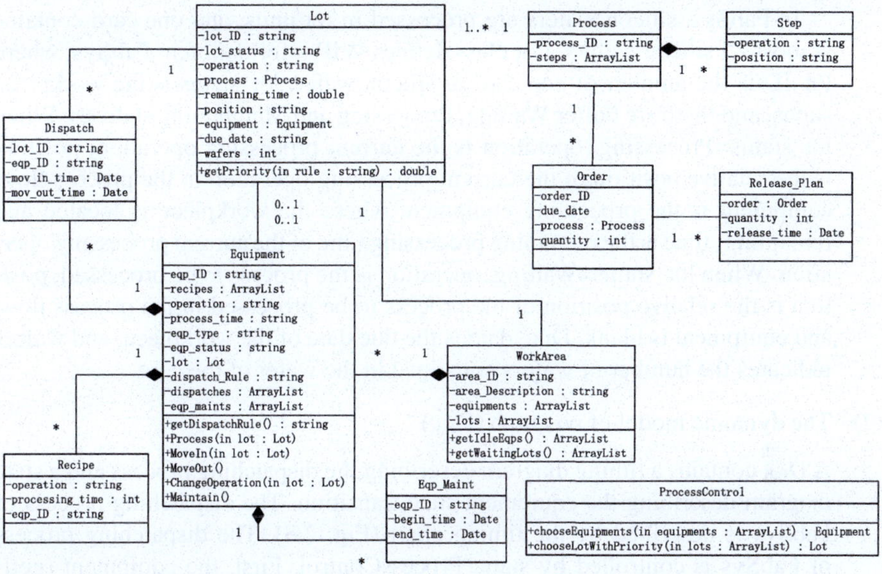

Fig. 2.7 Class diagram of Fabsys

processing capabilities between different equipment. Operation is the current machinable operation of the equipment. Processing_time is the processing time of the current operation. These two attributes represent the processing menu currently used by the equipment. Eqp_type is the equipment Processing type, which can be divided into three types according to the processing unit: lot processing, wafer processing and batch processing. eqp_status is the equipment status, and the equipment is Idle, Ready, and processing. Maintain four States. When eqp_status=Processing, the lot attribute indicates the silicon wafer being processed by the equipment; otherwise, the lot attribute is empty. Dispatch_rule indicates the scheduling rule used by the equipment when selecting the next silicon wafer to be processed, which is used to calculate the priority of the silicon wafer to be processed. Eqp_maints is a group of equipment maintenance plans, which are defined by Eqp_Maint class. In the time interval [begin_time, end_time], the number eqp_ID indicates that the equipment is in the maintenance period, so it is impossible to dispatch workers. Disptaches is a set of scheduling schemes, which are defined by the Dispatch class. In the time interval [mov_in_time, mov_out_time], the silicon wafer with lot_ID is processed on the equipment with eqp_ID, and both the silicon wafer corresponding to lot_ID and the equipment corresponding to eqp_ID are in Processing state. In FabSys, equipment is divided into different processing areas according to functions. The WorkArea class defines the processing areas, and the area_ID is the number of the processing areas. The processing areas contain equipment groups and lots of silicon wafers waiting to be processed in the buffer.

In FabSys, silicon wafers are processed in lot units, and one card contains at most 25 silicon wafers. Lot class defines WIP information in FabSys, where lot_ID is the number of one card of silicon wafer, lot_status is the workpiece status, and there are buffer Waiting, Processing and Maintaining statuses. When lot_status=Processing, operation is the current processing operation, position is the relative position of the current processing operation in the process flow, equipment is the processing equipment where the workpiece is located and remaining_time is the remaining processing time of the current processing operation. When lot_status=Waiting, operation is the process to be processed, position is the relative position of the process to be processed in the process flow, and equipment is blank. Due_date is the due date of the workpiece, and wafers indicates the number of wafers contained in the wafer of the card.

(2) The dynamic model of oosmfab (D_{fab})

A D_{fab} contains a timing diagram describing the dispatching process and a state diagram describing the equipment state transition. The dispatching process of FabSys is described by the timing chart (Fig. 2.8). The dispatching process of FabSys is controlled by static ProcessControl. First, the equipment (getIdleEqps ()) with Uchikoga in the processing area and no maintenance plan soon is obtained. When there is many idle equipment in the processing area, ChooseEquipments ()) by selecting the equipment with the least processing menu, and calculate the priority (getPriority ()) for each card waiting for the workpiece (getWaitingLots ()) in the buffer area of the processing area according to the scheduling rule (getDispatchRule ()) of the selected equipment. Choose the silicon wafer with the highest priority (chooseLotWithPriority ()) from the waiting silicon wafers and assign it to the selected equipment for processing (Process ()). The allocation processing is an asynchronous request, so you can continue to allocate silicon wafers for the remaining idle devices without waiting for the silicon wafer processing to be completed.

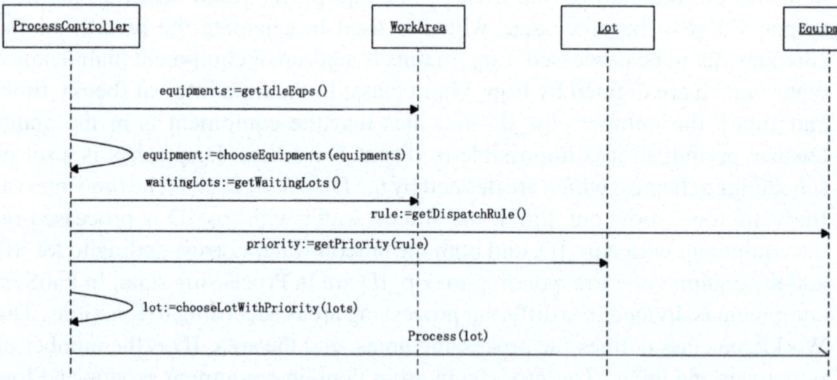

Fig. 2.8 Timing diagram of silicon wafer processing scheduling processes

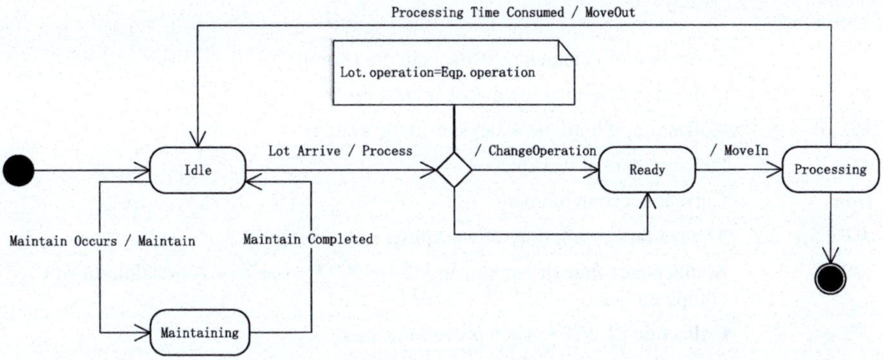

Fig. 2.9 State diagram of silicon wafer processed by equipment

The state diagram of the equipment (Fig. 2.9) shows the details of silicon wafer processing. When the silicon wafer reaches idle equipment, first check whether the current processing procedure of the equipment matches with the process to be processed (Lot.operation=Eqp.operation). If it does not match, it will enter the ready state directly. If it does not match, it will enter the ready state after changing the processing procedure (ChangeOperation()). When the equipment is ready, move the silicon wafer into the equipment (MoveIn ()) for processing until the processing time of this process is completed, and move the silicon wafer out of the equipment (MoveOut ()), and the equipment returns to the idle state again. When the equipment reaches the maintenance time, it will be (Maintain ()), and enter the maintenance state. It is impossible to dispatch the equipment in the maintenance state. When the maintenance is finished, it will be restored to the Idle state.

(3) Scheduling environment vector of Fabsys ($X_{se,fab}$)

The scheduling optimization problem of FabSys is a typical non-zero initial state scheduling problem. The scheduling environment of FabSys (such as the distribution of work-in-process in each processing area and the equipment status in each processing area, etc.) directly affects the results and performance of optimal scheduling. Scheduling environment of FabSys is described by vector $x_{se,fab}$. in Table 2.3, a set of variables are summarized to describe the scheduling environment of FabSys, where subscript X ∈ {5, 6} indicates that the silicon wafer model is 5 in. or 6 in., and subscript WA ∈ work_areas indicates the processing area. Components of $x_{se,fab}$ can be divided into production line scheduling environment variables and processing area scheduling environment variables. As shown in Tables 2.3 and 2.4.

First, define the parameters in Tables 2.3 and 2.4:

NL	Collection of work in process in system
NL_X	Collection of WIP of different categories in the system

(continued)

(continued)

NBL	Collection of emergency artifacts in the system
E	Collection of available devices in the system
BE	Collection of bottleneck devices in the system
D_i	Delivery date of workpiece i
Now	Current decision moment
$RPTS_{ij}$	The net processing time of workpiece i on equipment j
SDT_j	Maintenance time of equipment j ($24 - SDT_j$, one day's operation time of equipment j)
NL_{WA}	Collection of WIP in each processing area
NBL_{WA}	Emergency workpiece collection in each processing area
E_{WA}	Set of available equipment in each processing area
BE_{WA}	Collection of bottleneck equipment in each processing area

① Production line scheduling environment variables

Production line scheduling environment variables include current WIP quantity, WIP classification quantity, emergency workpiece quantity, emergency workpiece proportion, available equipment quantity, bottleneck equipment quantity, bottleneck equipment proportion, average remaining time of workpieces from current time to theoretical delivery time, standard deviation of remaining time of workpieces from current time to theoretical delivery time and system processing capacity ratio.

② The processing area scheduling environment variables

The following attributes are considered in the scheduling environment variables of processing zones: the number of WIP in each processing zone, the ratio of WIP in each processing zone to total WIP, the ratio of processing capacity in each processing zone, the number of available equipment in each processing zone, the number of bottleneck equipment in each processing zone, and the ratio of bottleneck equipment in each processing zone to available equipment in this area.

(4) The scheduling method module (candidate_rule) of Fabsys and the scheduling method setting coding rule ($X_{ruleset}$)

According to the dynamic model D_{fab} of $OOSM_{fab}$, FabSys selects the workpiece with the highest priority and assigns it to idle equipment for processing according to the corresponding scheduling rules, to generate a scheduling scheme and optimize the scheduling performance. In different scheduling environments, different scheduling rules are set according to different scheduling objectives. At the same time, the equipment groups have different scheduling rule bases to choose from because of their different process characteristics. Whether the scheduling strategy can be reasonably selected has an important impact on

Table 2.3 Production line scheduling environment variables

Attribute name	Attribute meaning	Mathematical description
WIP	Current WIP quantity in the system	$\|NL\|$
WIP X	Classification quantity of WIP in the system	$\|NL_X\|$
NoBL	Number of emergency workpieces in the system	$\|NBL\|$
PoBL	Proportion of emergency workpieces in the system	$\|NBL\|/\|NL\|$
NoE	Number of devices currently available in the system	$\|E\|$
NoBE	Number of bottleneck devices in the system	$\|BE\|$
PoBE	Proportion of bottleneck equipment in the system	$\|BE\|/\|E\|$
MeTD	The average remaining time of the workpiece in the system from the current time to the theoretical delivery time	$\left(\sum_{i\in NL}\|D_i - Now\|\right)/\|NL\|$
SdTD	The standard deviation of the remaining time of the workpiece from the current time to the theoretical delivery date in the system	$\sqrt{\left(\sum_{i\in NL}[(D_i - Now) - MeTD]^2/\|NL\|\right)}$
PC	System processing capacity ratio	$\sum_{i\in NL, j\in E} PRTS_{ij}/\sum_{j\in E}(24 - SDT_j)$

the performance index of the production line after the end of the production scheduling cycle. OOSM_{fab} sets real-time scheduling rules according to processing area, making candidate_rule WA an optional scheduling rule set of processing area WA, and $X_{\text{rule DF}} \in$ candidate_rule WA represents the scheduling rule adopted by processing area WA. Rule setting vector xruleset = ($X_{\text{rule Df}}$, $X_{\text{rule IM}}$, $X_{\text{rule EP}}$, $X_{\text{rule LT}}$, $X_{\text{rule PE}}$, $X_{\text{rule PC}}$, $X_{\text{rule TF}}$, $X_{\text{rule WT}}$) indicates the setting of real-time scheduling rules for each processing area. The real-time scheduling rule set (candidate_rule) is given in Table 2.5. Candidate_rule is implemented in OOSM_{fab} in the way of Method.

In addition, the releasing strategy of the order will also have an impact on the performance index, which is recorded as release. Because this paper mainly studies the scheduling problem in FabSys short scheduling period, the releasing strategy is not considered in X_{ruleset}. The releasing strategy adopts Constant WIP, CONWIP) by default, that is, release = CONWIP.

Table 2.4 Scheduling environment variables in processing zones

Attribute name	Attribute meaning	Mathematical description
WIP WA	Quantity of work-in-process in processing area WA	$\lvert NL_{WA}\rvert$
PoB WA	The proportion of WIP in the total WIP in the processing zone WA	$\lvert NL_{WA}\rvert/\lvert NL\rvert$
NoBL WA	Emergency workpiece quantity of WA in processing area	$\lvert NBL_{WA}\rvert$
PoBL WA	Proportion of urgent workpieces in WA processing area	$\lvert NBL_{WA}\rvert/\lvert NL_{WA}\rvert$
PC WA	Production capacity ratio of WA processing in processing area	$\sum_{i\in NLWA, j\in EWA} RPTS_{ij}/\sum_{j\in EWA}(24 - SDT_j)$
NoE WA	Available equipment quantity in processing area WA	$\lvert E_{WA}\rvert$
NoBE WA	Number of WA Bottleneck Equipment in Processing Zone	$\lvert BE_{WA}\rvert$
PoBE WA	Proportion of WA bottleneck equipment in processing area to available equipment in this area	$\lvert BE_{WA}\rvert/\lvert E_{WA}\rvert$

Parameters used in Table 2.5 are defined as follows:

P_i	Scheduling priority of artifact i
D_i	Delivery date of workpiece i
F_i	Production cycle multiplication factor of workpiece i
Q_i	A target WIP value of the product to which the workpiece i belongs
N_i	Current WIP value of products to which workpiece i belongs
PT_{in}	Time for processing the nth process of workpiece i, including waiting time
AT_i	The moment when the workpiece i enters the buffer zone
CR_{ik}	Critical value when workpiece i is to be processed in the kth process
OD_{ik}	Decision value when the workpiece i will be processed in the kth process
RP_i	Planned remaining machinable time of workpiece i
NQ_i	Workpiece i the number of workpieces to be processed before the equipment in the next process of the process to be processed
Now	Current decision moment
AWT_{ik}	The waiting time after the processing of the workpiece i completes the kth process
SPT_i	Entering time of workpiece i

(continued)

(continued)

RPT_{ik}	Total processing time of workpiece i currently used, including waiting time
$TRPT_{ik}$	The remaining net processing time after the kth process of workpiece i
$ProTime_{ik}$	Machining time of the current process for machining the workpiece i on the equipment k

(5) Scheduling performance index of Fabsys (P_{fab})

The scheduling performance index of FabSys is the evaluation basis of FabSys scheduling scheme, which can be divided into two categories, one is short-term performance index, such as WIP quantity, total moving quantity, average moving quantity, and equipment utilization rate, and the other is long-term performance index, such as average processing cycle and on-time delivery rate. Specific definitions are as follows:

Work-in-process level (WIP): the number of all unfinished workpieces on the production line. The level of work-in-process in the production line should be consistent with the expected target as far as possible. Too little will make the equipment idle, which will not make good use of the production capacity, while too much will lead to the lengthening of the processing cycle and affect the delivery time.

Productivity (Prod): the number of workpieces completed by the production line in unit time. The higher the productivity, the more workpieces are completed in unit time, and the higher the utilization rate of equipment, which is helpful to shorten the processing cycle.

Machining Cycle Time (CT): the time taken for an original workpiece to enter the machining system and leave the machining system as a finished product.

Machine Utility (Utility): the ratio of the time when the equipment is in processing state to its startup time. The equipment utilization rate is related to the WIP quantity, and the higher the WIP quantity, the higher the equipment utilization rate. However, when the WIP quantity is saturated, if the WIP quantity is increased, the equipment utilization rate will not increase.

Total Movement (MOV): the total number of steps that all workpieces move in unit time. The higher the total movement, the higher the number of processing tasks completed by the production line, the more the total movement of the production line, and the higher the equipment utilization rate.

Average Turn: the average number of moving steps of a workpiece per unit time. The higher the moving speed, the faster the flow speed of the production line, which helps to shorten the average processing cycle.

On-time Delivery Rate (ODR): the percentage of workpieces delivered on time to finished workpieces.

Set the above performance indicators as P_{fab}.

(6) Functional model of Fabsys (F_{fab})

To sum up, scheduling environment (X_{se}), scheduling method setting code ($X_{ruleset}$) and definition of performance index (P_{fab}) make scheduling period

Table 2.5 Real-time scheduling rules

Rule name	Rule description	Mathematical description
The earliest delivery date is preferred (earlies due date, EDD)	It is the workpiece with the earliest delivery date that will be processed first	$D_i < D_j (i \neq j) \Rightarrow P_i > P_j$
The earliest process delivery time is preferred (earlies operation due date, EODD)	Workpieces with the earliest process delivery time will be processed first. The process delivery time of a workpiece can be determined by the incoming time of the workpiece, the total processing time currently used and the multiplication factor of the production cycle	$OD_{im} < OD_{jn} (i \neq j) \Rightarrow P_i > P_j$ $OD_{ik} = SPT_i + RPT_{ik} * F_i$ $RPT_{ik} = \sum_{n=1}^{k} PT_{in}$
Minimum critical value first (critical ratio, CR)	The processing sequence of workpieces is sorted based on the due date, current time, and the remaining net processing time of the workpieces	$CR_{im} < CR_{jn} (i \neq j) \Rightarrow P_i < P_j$ $CR_{ik} =$ $(1 + TRPT_{ik})/(1 + D_i - Now)$; $Now < D_i$
The longest processing time, LPT first (LPT)	The current process of the workpiece takes up the shortest equipment time and is processed first	$ProTime_{im} < ProTime_{jn} (i \neq j) \Rightarrow P_i > P_j$
Shortest remaining processing time (SRPT)	The workpiece with the shortest remaining processing time will be processed first	$RP_i < RP_j (i \neq j) \Rightarrow P_i > P_j$; $RP_i =$ $D_i - Now$
First in first out (FIFO) (first in first out, FIFO)	Workpieces that arrive at the buffer zone first are processed first	$AT_i < AT_j (i \neq j) \Rightarrow P_i > P_j$
Shortest waiting time first (list scheduling, LS)	The workpiece with the shortest waiting time can be processed preferentially. The available waiting time for processing is determined by the due date of the workpiece, the remaining net processing time, and the current decision moment	$AWT_{im} < AWT_{jn} (i \neq j) \Rightarrow P_i > P_j$ $AWT_{ik} = D_i - TRPT_{ik} - Now$

(continued)

Table 2.5 (continued)

Rule name	Rule description	Mathematical description
Fewest lots at the next queue, FLNQ)	The workpiece with the smallest next queue is processed first. The next queue of workpieces refers to the number of workpieces waiting to be processed before the equipment in the next process of the workpiece to be processed	$NQ_i < NQ_j (i \neq j) \Rightarrow P_i > P_j$
Load (load balance, LB)	Make those workpieces with large deviation from the established WIP target have higher priority	$\sum_{i \in NL} \lvert D_i - Now \rvert / \lvert NL \rvert$
Blanket rules (general rule, GR)	The scheduling rules applied by FabSys consider many factors, such as process constraints, delivery time, customer priority, remaining processes and so on	

T = 12 h, and it is easy to get the functional model of $OOSM_{fab}$.

$$F_{fab} = \{Y_{pi} = fpi(X_{se,fab}, X_{ruleset}) | p_i \in P_{fab}\}.$$

(7) Data model of Fabsys

Online data can be used to construct the data model of FabSys, to obtain the information needed for production line modeling. Online static data reflects the static attributes of FabSys, process flow information and product specification information define the processing path and product type of workpieces respectively, and equipment processing capacity information and processing area layout information define the processing capacity and equipment grouping layout of production line. Online dynamic data reflects the scheduling environment of FabSys, including workpiece status and equipment status. With ORM mapping, online data model can be dynamically loaded, and the object model of simulation model $OOSM_{fab}$ can be constructed. Online data model is defined by Table 2.6, the variable values in FabSys's scheduling environment set $X_{se,\ fab}$ can be obtained from the data defined in Table 2.6.

$OOSM_{fab}$ can generate the releasing plan and dispatching scheme as shown in Table 2.7 by setting real-time scheduling rules and production planning strategies and operating through simulation models. The performance index values defined in the performance index set P_{fab} of $OOSM_{fab}$ operation results can be

Table 2.6 Online data model of Fabsys

Data type	Information type	Data sheet	Data attribute name	Physical significance	Whether primary key	Whether foreign key
Static state data	Process process Information	Process	Process_ID	Routing number	√	
			Operation	Operation name	√	
			Position	Position of operation in process		
	Products order information	Order	Order_ID	Order number	√	
			Due_Date	Order lead time		
			Quantity	Order demand artifact quantity		
			Process_ID	Routing number of the order		√
	Equipment processing competence information	Equipment	Eqp_ID	Equipment number	√	
			Eqp_Type	Equipment processing type		
			Area_ID	No. of the processing zone where the equipment is located		√
		Recipe	Eqp_ID	Equipment number	√	
			Operation	Operation name	√	
			Process_Time	Processing time		
	Layout information	WorkArea	Area_ID	Processing area number	√	
			Area_Description	Description of processing area		

(continued)

Table 2.6 (continued)

Data type	Information type	Data sheet	Data attribute name	Physical significance	Whether primary key	Whether foreign key
Dynamic data	Equipment state information	Equipment	Eqp_Status	Device status {processing, ready, idle}		
			Operation	Equipment current operation		
			Process_Time	Processing time of current operation of equipment		
			Dispatch_Rule	Scheduling rules of equipment		
			Lot_ID	Workpiece number being processed by the equipment		√
		Eqp_Maint	Eqp_ID	Equipment number to be maintained	√	
			Begin_Time	Equipment maintenance start time		
			End_Time	End time of equipment maintenance		
	Workpiece state information	Lot	Lot_ID	Workpiece number	√	
			Lot_Status	Workpiece status {processing, waiting}		
			Eqp_ID	Number of processing equipment where the workpiece is located		√

(continued)

Table 2.6 (continued)

Data type	Information type	Data sheet	Data attribute name	Physical significance	Whether primary key	Whether foreign key
			Area_ID	No. of the processing area where the workpiece is located		√
			Position	The current processing operation position, if the workpiece is waiting, is the operation position to be processed		√
			Operation	Current processing operation name, if the workpiece is waiting, it is the operation name to be processed		
			Process_ID	Routing number		√
			Order_ID	Order number		√

obtained from the data defined in Table 2.7. The entity relationship diagram of the data models defined in Tables 2.6 and 2.7 is shown in Fig. 2.10.

Offline historical data of FabSys records the actual operation process of FabSys. From these historical data, uncertain parameters such as equipment processing time, uncertain events such as equipment maintenance, equipment failure and urgent orders, and prediction models of performance indicators such as CT and WIP can be extracted, thus further improving the modeling accuracy and scheduling effect of $OOSM_{fab}$. Table 2.8 lists the offline historical data that FabSys can use to construct uncertain parameters and event prediction models.

(8) Analysis of historical status data of Fabsys, taking the distribution of work in process as an example

In order to analyze the statistical characteristics of variables in X SE and FAB, the distribution of WIP in each processing area is analyzed. Sampling

Table 2.7 $OOSM_{fab}$ generated data

Information type	Data sheet	Data attribute name	Physical significance
Dispatch scheme	Dispatch	Eqp_ID	Equipment number
		Lot_ID	Equipment processing procedure
		Move_In_Time	Entry time of workpiece into equipment
		Move_Out_Time	Workpiece removal time from equipment
Releasing plan	Release_Plan	Order_ID	Equipment number
		Quantity	Equipment processing procedure
		Release_Time	Entry time of workpiece into equipment

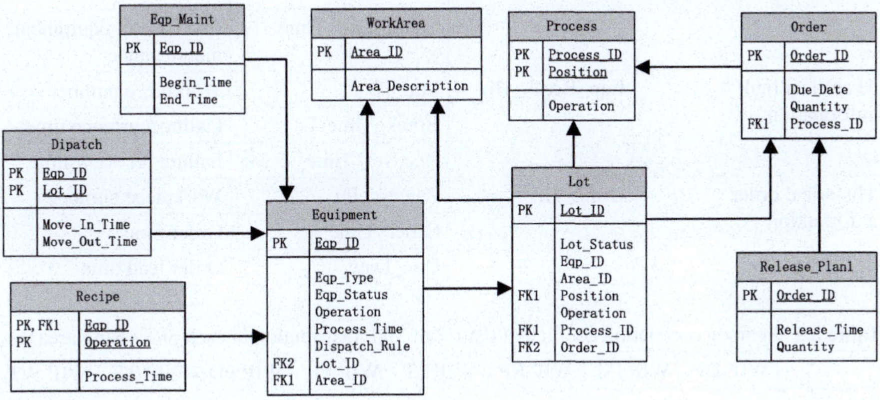

Fig. 2.10 Entity relationship diagram of $OOSM_{Fab}$ data model

the Online dynamic data of MES system of FabSys from January 1, 2012 to May 10, 2012 once every four hours, extracting WIP WA of WIP distribution in each processing area, calculating Pearson correlation coefficient matrix and Kolmogorov–Smirnov test of WIP distribution in each processing area, and obtaining the results shown in Tables 2.9 and 2.10.

From the results in Table 2.9, it can be seen that there is a strong linear coupling between WIP distributions in processing areas, especially between upstream and downstream processing areas. The coupling between WIP distributions in lithography areas and WIP distributions in other processing areas is particularly prominent. According to the inspection results in Table 2.10, the p value of WIP distribution variables in other processing areas except WIP TF is less than 0.05, so these WIP distributions do not obey normal distribution. Although WIP TF obeys normal distribution, its P value is close to 0.05, and

Table 2.8 Offline historical data of Fabsys

Information type	Data sheet	Data attribute name	Physical significance
Historical machining information	Lot_Move_His	Eqp_ID	Equipment number
		Operation	Equipment processing procedure
		Process_ID	Workpiece processing technology number
		Position	Current machining operation position
		Move_In_Time	Entry time of workpiece into equipment
		Move_Out_Time	Workpiece removal time from equipment
Historical maintenance information	Eqp_Maint_His	Eqp_ID	Equipment number
		Maint_Begin_Time	Equipment maintenance start time
		Maint_End_Time	End time of equipment maintenance
Historical fault information	Eqp_Break_His	Eqp_ID	Workpiece number
		Break_Time	Fault occurrence time
		Recover_Time	Failure recovery time
Historical order information	Order_His	Process_ID	Workpiece number
		Order_Time	Order time
		Due_Date	Order lead time

Table 2.9 Pearson correlation coefficient matrix of WIP distribution in each processing area

	WIP DF	WIP IM	WIP EP	WIP LT	WIP PE	WIP PD	WIP TF	WIP WT
WIP DF	1	0.47	0.29	0.08	0.28	− 0.08	− 0.12	0.40
WIP IM	0.47	1	− 0.22	− 0.36	0.16	− 0.10	− 0.25	0.16
WIP EP	0.29	− 0.22	1	0.83	− 0.01	− 0.27	− 0.07	0.69
WIP LT	0.08	− 0.36	0.83	1	− 0.32	− 0.44	− 0.16	0.63
WIP PE	0.28	0.16	0.01	− 0.32	1	0.35	− 0.19	− 0.12
WIP PD	− 0.08	− 0.10	− 0.27	− 0.44	0.35	1	0.07	− 0.35
WIP TF	− 0.12	− 0.25	− 0.07	− 0.16	− 0.19	0.07	1	0.04
WIP WT	0.41	0.16	0.69	0.63	− 0.12	− 0.35	0.04	1

Table 2.10 Kolmogorov–Smirnov test of WIP distribution in each processing area

	WIP DF	WIP IM	WIP EP	WIP LT	WIP PE	WIP PD	WIP TF	WIP WT
P value	< 0.05	< 0.05	< 0.05	< 0.05	< 0.05	< 0.05	0.07	< 0.05

the confidence of WIP TF obeys normal distribution is not high. On the one hand, the reason for this result is the inherent complexity of FabSys manufacturing process, and on the other hand, the data in enterprise information system contains noise due to manual operation errors and other reasons.

To preprocess and mine the data with high coupling and complex distribution, this paper proposes a data preprocessing and data modeling method based on computational intelligence, which improves the data preprocessing quality and generalization ability of data mining by iterative method.

2.3.3 Data-Driven Forecasting Model in FabSys

(1) FabSys data-driven model parameter prediction model

The FabSys data-driven model parameter prediction model is directly learned from offline historical data of FabSys and driven by online dynamic data. Taking the processing time of silicon wafer as an example, the coefficient α_i of the ith influencing factor (which α_0 is a constant term) can be obtained by using the processing history records to construct a linear regression model (2.1). In which, $PT_t^{eqp\_id,op}$ is the estimated processing time of equipment with equipment number eqp_id and current processing procedure is op, duration op is the duration for equipment to keep current processing procedure (if duration op = 0, setting time of procedure switching should also be considered), lot.wafer_count is the number of silicon wafers contained in one card of silicon wafers currently processed, and, $PT_{t-1}^{eqp\_id,op}$ $PT_{t-2}^{eqp\_id,op}$ $PT_{t-3}^{eqp\_id,op}$, is the time consumed by the first three processing procedures op of equipment with equipment number eqp_id. Maintain the duration of the current processing operation for the equipment.

$$PT_t^{eqp_{id},op} = \alpha_0 + \alpha_1 * duration_{op} + \alpha_2 * lot.wafer_{count} + \alpha_3 * PT_{t-1}^{eqp_{id},op} + \alpha_4 * PT_{t-2}^{eqp\_id,op} + \alpha_5 * PT_{t-3}^{eqp\_id,op} \quad (2.1)$$

Uncertain parameters and events in $OOSM_{fab}$ are predicted before $OOSM_{fab}$ runs, which can improve the accuracy of $OOSM_{fab}$ running results. The modeling method of FabSys data-driven model parameter prediction is shown in Fig. 2.11.

(2) FabSys data-driven performance prediction model

For a large-scale complex manufacturing system like FabSys, it is a time-consuming process to obtain performance indicators through $OOSM_{fab}$ online simulation. Data-based performance index prediction modeling method can respond quickly and obtain the predicted value of performance index. Therefore, in the model layer, a performance index prediction model based on data is introduced. The overall situation is shown in Fig. 2.11. The performance index

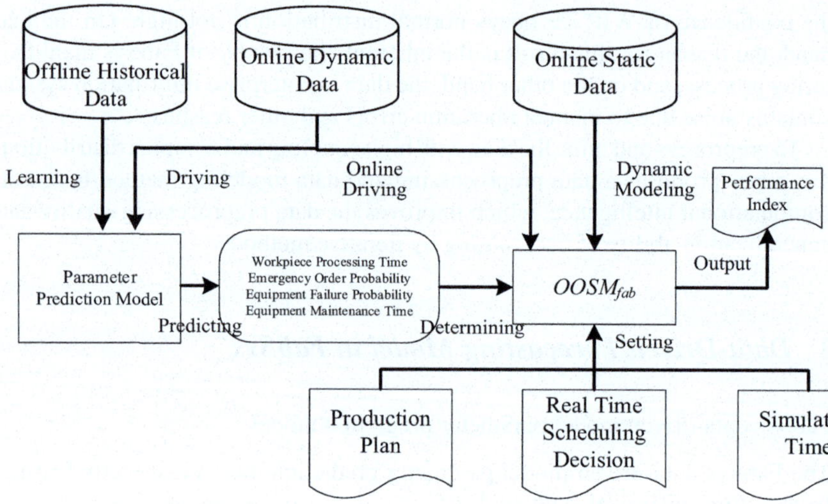

Fig. 2.11 FabSys data-driven parameter prediction model construction method

prediction model can be obtained by generating offline performance index simulation data through many offline simulations, and then mining data from these data.

Performance index prediction models can be divided into global performance index and local performance index prediction models. According to the simulation time (or forecast cycle), it can be divided into real-time (forecast cycle is recorded in hours), short-term (forecast cycle is recorded in diary) and long-term (forecast cycle is recorded in weeks) performance index prediction models. When the forecast period is in hours, the global performance index changes little, and the local short-term performance index forecast model is mainly concerned. When the forecast cycle is written weekly, the global long-term performance indicators are mainly concerned. When the forecast period is a diary, it is necessary to consider both global short-term performance indicators and local short-term performance indicators. In addition, for different forecasting models, the influencing factors are also different. For forecasting models with daily global performance indicators, the values of $_{se,fab}$ and $X_{ruleset}$ should be considered. For long-term global performance indicators forecasting models with weekly forecast periods, uncertain factors in manufacturing systems and release of releasing strategy should also be considered. For the local performance index prediction model with short prediction period, the influencing factors can be obtained by selecting several dimensions from $X_{se,fab}$ and $X_{ruleset}$ through feature selection algorithm.

Figure 2.12 takes the performance index prediction model with daily prediction cycle as an example to predict the average utilization rate of manufacturing system equipment. The main influencing factors are the initial scheduling environment of FabSys, namely X_{se}, the value of Fab, and the value of $X_{ruleset}$,

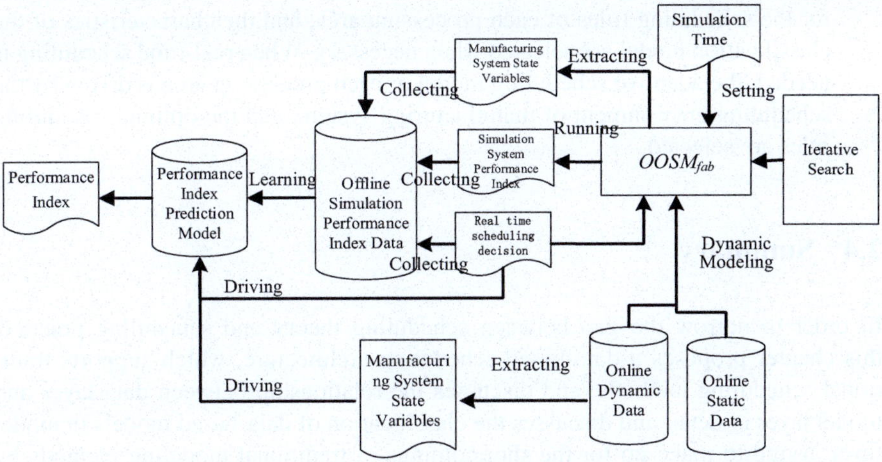

Fig. 2.12 Construction method of Fabsys data-driven performance index prediction model

which is the scheduling rule adopted by each processing area. We can learn the performance index prediction model f Utility from many offline simulation performance indexes generated by offline simulation and get the predicted value of Utility

$$Y_{Utility} = f'_{Utility}(X_{se,fab}, X_{ruleset}) \tag{2.2}$$

(3) FabSys data-driven adaptive scheduling model

Due to the large scale of FabSys, it is very time-consuming to select the optimized scheduling scheme through online optimization. In order to make fast optimal scheduling decisions for the performance indicators that need to be optimized online, offline optimization can be used to optimize the performance indicators, generate offline simulation optimization scheduling decision data, mine the data and construct an adaptive scheduling model. The dispatching decision can be made directly corresponding to the scheduling performance index to be optimized according to the current scheduling environment.

Different from the performance index prediction problem, the adaptive scheduling model optimizes the performance index in the offline optimization stage, and the optimization target can be single performance index or multiple performance indexes. In FabSys, the scheduling rules of each processing area are coded, and the performance index is optimized by exhaustive search or heuristic search. The optimized scheduling rule combination of each processing area is obtained and saved as the Offline simulation optimization scheduling decision data. Because the optimal scheduling scheme is a decision combination, the final adaptive optimal scheduling decision problem can be decomposed into several classification problems, that is, the classification model is constructed for the scheduling rules of each processing area, and the characteristics of the classification model are selected when necessary. When real-time scheduling is needed, the adaptive scheduling model of each processing area is driven by the scheduling environment of manufacturing system, and the optimal scheduling rules are selected.

2.4 Summary

In order to narrow the gap between scheduling theory and scheduling practice, this chapter proposes a data-based scheduling architecture, which supports traditional scheduling methods, and discusses the relationship between data layer and model layer in detail and discusses the classification of data-based models in model layer, trying to make up for the shortcomings of traditional modeling methods by means of data-based modeling. A complex silicon wafer processing system FabSys is introduced, and its solution based on data scheduling architecture is designed.

References

1. Li Q, Chen Y (2004) Overall design of enterprise informatization. Tsinghua University Publishing House
2. Pandey D, Kulkarni MS, Vrat P (2010) Joint consideration of production scheduling, maintenance and quality policies: a review and conceptual framework. Int J Adv Oper Manag 2(1):1–24
3. Monfared MAS, Yang JB (2007) Design of integrated manufacturing planning, scheduling and control systems: a new framework for automation. Int J Adv Manuf Technol 33(5–6):545–559
4. Wang F, Chua TJ, Liu W et al (2005) An integrated modeling framework for capacity planning and production scheduling. In: International conference on control and automation, 2005. ICCA'05, vol 2. IEEE, pp 1137–1142
5. Lalas C, Mourtzis D, Papakostas N et al (2006) A simulation-based hybrid backwards scheduling framework for manufacturing systems. Int J Comput Integr Manuf 19(8):762–774
6. Lin JT, Chen TL, Lin YT (2006) A hierarchical planning and scheduling framework for TFT-LCD production chain. In: IEEE international conference on service operations and logistics, and informatics, 2006. SOLI'06. IEEE, pp 711–716
7. Framinan JM, Ruiz R (2010) Architecture of manufacturing scheduling systems: literature review and an integrated proposal. Eur J Oper Res 205(2):237–246
8. Sadeh NM, Hildum DW, Laliberty TJ et al (1998) A blackboard architecture for integrating process planning and production scheduling. Concurr Eng 6(2):88–100
9. Nishioka Y (2004) Collaborative agents for production planning and scheduling (CAPPS): a challenge to develop a new software system architecture for manufacturing management in Japan. Int J Prod Res 42(17):3355–3368
10. Gómez-Gasquet P, Lario FC, Franco RD et al (2011) A framework for improving Planning-Scheduling Collaboration in industrial production environment. Stud Inform Control 20(1):68
11. Tai TT, Boucher TO (2002) An architecture for scheduling and control in flexible manufacturing systems using distributed objects. IEEE Trans Robot Autom 18(4):452–462
12. Li LI, Fei Q (2012) A modular simulation system for semiconductor manufacturing scheduling. Przegl ą d Elektrotechniczny 88(1b):12–18

13. Govind N, Bullock EW, He L et al (2008) Operations management in automated semiconductor manufacturing with integrated targeting, near real-time scheduling, and dispatching. IEEE Trans Semicond Manuf 21(3):363–370
14. Niu L, Zhou H, Jia S et al (2009) Research on the design of job scheduling system model base based on Unified Modeling Language. Comput Integr Manuf Syst 15(3):451–457
15. Qiao F, Ma YM, Li L et al (2013) A Petri net and extended genetic algorithm combined scheduling method for wafer fabrication

Chapter 3
Data Preprocessing of Semiconductor Manufacturing System

The real-world data is disturbed by various factors, and its quality is not high. Low-quality data will lead to poor mining results, so data preprocessing is usually regarded as an important part of data mining. The purpose of data preprocessing is to improve data quality, which generally includes data integration, transformation, cleaning, and reduction. In the process of data acquisition in semiconductor manufacturing system, sensor drift, equipment failure, or operator input error will inevitably occur, resulting in noise in the data set. In addition, the data related to production scheduling needs to be integrated from MES, ERP, SCADA, and other systems. The data in these systems describe the enterprise production process from different levels and angles, resulting in high redundancy between the attributes of the integrated data, which requires data preprocessing.

3.1 Introduction

With the development of modern industrial technology, the manufacturing process, process, and equipment tend to be complicated, and it is difficult to accurately model the system by the traditional modeling method of mechanism model to optimize the system performance. For example, for the silicon wafer processing production line [1], although the advanced scheduling idea is applied, the scheduling algorithm is carefully designed and implemented, but the simulation results obtained have poor accuracy, which is difficult to guide the actual scheduling and scheduling tasks [2, 3]. With the improvement of enterprise informationization, the real-time and accuracy of data collected by manufacturing enterprises have been significantly improved, thus promoting the application of data-based methods in manufacturing process control, online monitoring, and fault diagnosis, scheduling optimization and management decision optimization [4, 5]. Especially in the field of iron and steel metallurgy, because its key performance indicators can not be described by the mechanism model

L. Li et al., *Data-Driven Scheduling of Semiconductor Manufacturing Systems*,
Advanced and Intelligent Manufacturing in China,
https://doi.org/10.1007/978-981-19-7588-2_3

or monitored online, the data-based prediction method has been widely used [6–8]. Data-based scheduling method focuses on the combination of data-driven method and traditional scheduling modeling optimization method to solve scheduling problems. This section will explain three aspects: complex manufacturing data attribute selection, complex manufacturing data clustering and complex manufacturing data attribute discretization.

(1) Attribute selection of complex manufacturing data

Excessive redundancy of conditional attributes will lead to the decline of classification or regression accuracy, resulting in unusable generated rules and more conflicts between rules. Attribute selection is to select more important attributes from conditional attributes. Common methods of attribute selection include rough set, computational intelligence and so on. For example, aiming at the quality problems of semiconductor manufacturing, Kusiak [9–11] proposed a method of obtaining rules from sample data based on rough set, and applied feature transformation and data set decomposition technology to improve the accuracy and efficiency of defect prediction; Attribute reduction of rough set is an NP-hard problem. Chen and others [12] reduced the search space by the concept of feature kernel, and then used ant colony algorithm to obtain attribute set reduction, which improved the efficiency of knowledge reduction. Shiue and others [13–17] have established a two-stage decision tree adaptive scheduling system, which uses a neural network-based weight feature selection algorithm and genetic algorithm for scheduling attribute selection, uses Self-Organizing Maps (SOM) for data clustering, and uses three learning algorithms, decision tree, neural network and support vector machine, to learn each cluster to realize parameter optimization, thus improving the generalization ability of the adaptive scheduling knowledge base, and verifying the effectiveness of the results through simulation.

(2) Complex manufacturing data clustering

Clustering is a technique to classify sample data according to their similarity, so that similar samples belong to the same class, while samples with low similarity belong to different classes. Noise data will affect the accuracy of learning. For example, when dealing with noisy samples, C4.5 will cause the spanning tree to be large, which will reduce the prediction accuracy and require pruning. Therefore, for large-scale training samples, clustering can be used to smooth the noise data. Commonly used methods in clustering include SOM, Fuzzy-C mean, K mean, neural network and so on. For example, Hu and Su [18] use hierarchical clustering to find out the equipment related to the decline in yield; Chen [19, 20] used Fuzzy-C mean and k mean to cluster the training samples, and then trained neural network for each cluster to improve the prediction accuracy of workpiece processing cycle.

(3) Attribute discretization of complex manufacturing data

Some algorithms and models can only deal with discrete data, such as decision tree and rough set, so it is necessary to use attribute discretization technology to

Table 3.1 Data preprocessing tasks of manufacturing system

Data preprocessing task	Solution method
Data conversion	Data normalization
Data cleanup	Missing value filling and outlier detection
Data specification	Redundant variable detection

transform continuous attribute values into discrete attribute values. For example, when Knooce [21] and Li [22] mine the optimal scheduling scheme, according to the characteristics of attribute-oriented reduction algorithm and decision tree, the attribute values are divided into equidistant discrete partitions; Rafinejad [23] proposed an attribute discretization method based on fuzzy K-means algorithm, which makes the rules extracted from the optimal scheduling scheme approach the optimal scheduling scheme better.

The existing preprocessing techniques of complex manufacturing mainly focus on attribute selection and data clustering, while the preprocessing techniques of manufacturing system data have the characteristics of large scale, noise, complex sample distribution and missing phenomenon, large number and variety of input variables, nonlinear and strong coupling among input/output variables, etc., which need to be further studied. In this chapter, aiming at the noisy and highly redundant production scheduling data, the problems such as data normalization, missing value filling, outlier detection and redundant variable detection are extracted corresponding to the data preprocessing tasks, as shown in Table 3.1, and the solutions to these problems are given, as shown in Fig. 3.1. The methods belong to PreProcData in DataProcAnalyModule in DSACMS.

For data-based scheduling prediction modeling problems (such as scheduling parameter prediction), it is necessary to obtain relevant data from multiple heterogeneous data sources, that is, ETL defined in DataProcAnalyModule of DSACMS. The information system of object production line uses relational database to store

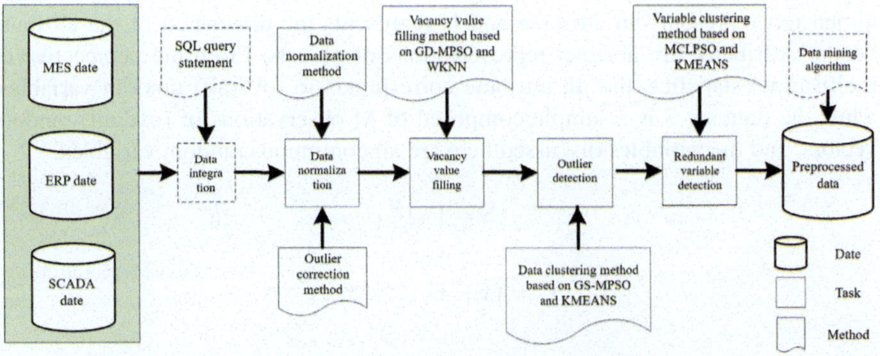

Fig. 3.1 Technical route of data preprocessing in manufacturing system

data, so data integration can be realized by Structured Query Language, SQL). For the integrated data, it is necessary to transform it into a convenient data mining form. The following sections will introduce the methods respectively. Section 3.2 introduces the methods of variable normalization and outlier correction for smoothing data noise commonly used in statistical analysis. Section 3.3 discusses the missing value filling technology in data cleaning and proposes a weighted attributes k nearest neighbors (WAKNN) filling method based on GD-MPSO to deal with the incomplete data caused by data noise. Section 3.4 gives an outlier detection method based on data clustering based on GS-PSO and K-means clustering method, which can detect outliers in data set and realize data cleaning. Section 3.5 presents a redundant variable detection method based on variable clustering based on MCLPSO and Principal Component Analysis, PCA), which can detect redundant attributes in data sets and realize data reduction.

In this chapter, two data sets collected from actual manufacturing information system will be used to verify the above method. Data set D_1 is the scheduling environment data collected from MES of FabSys. The scheduling environment is described by variables in $X_{SE,Fab}$, including 67 status attributes, 542 sample data from January 1, 2012, to May 2, 2012. D_2 is the machine learning public test data set provided by UCI (University of California Irvine), and D_2 is the sensor data collected from the monitoring system of a semiconductor production line. The original data includes 591 attributes representing sensors and 1567 sample data from July 19, 2008, to October 15, 2008. After data cleaning operations (1)–(3), the data in D_2 includes 440 sensors and 1561 sample data.

(1) Delete invalid sensor: the value of the sensor is constant, and the missing value ratio of data collected by the sensor $\geq 50\%$.
(2) Delete the sample data with more vacant values: $\geq 30\%$ of the sensor attribute values in the sample data are vacant.
(3) The remaining missing values are filled with the sensor mean value.

For the convenience of discussion, the data set in this paper is defined as follows: Data set S is a set composed of m records, in which records describe a specific object, which is usually represented by an N-dimensional attribute vector, where each dimension represents an attribute and N represents the dimension of the attribute vector. Attributes are abstract representations of objects. From the perspective of multivariate statistics, the ith attribute corresponds to (overall) random variables, while the data set S is a sample composed of M observations of (overall) random vectors, and the variables discussed here are all continuous random variables.

$$\mathrm{S} = \{x_i\}_{i=1}^{M}$$

$$x_i = (x_{i1}, x_{i2}, \ldots, x_{iN})$$

$$X = (X_1, X_2, \ldots, X_N)$$

3.2 Data Standardization

3.2.1 Data Normalization Rules

Data normalization refers to scaling attribute data of data set S according to rules, so that it falls into a specific interval. Data normalization can eliminate the influence of dimensional differences of different attributes on data analysis results. Practice has proved that for multilayer perceptron neural network using back propagation learning algorithm, standardizing the input value of measuring each attribute in training tuple is helpful to speed up the learning stage; For K-means clustering, data normalization can make all attributes have the same weight. Therefore, data normalization is a necessary preparation step for data analysis. This section introduces the two most used data normalization methods [24], maximum and minimum normalization and z-score normalization.

(1) Maximum and minimum standardization

$$x'_{li} = \frac{x_{li} - \min_{Xi}}{\max_{Xi} - \min_{Xi}}\left(new\_\max_{Xi} - new\_\min_{Xi}\right) + new\_\min_{Xi} \tag{3.1}$$

where x_{li} is the lth observed value of the variable X_i. That is, the value of the attribute i of the record l in the data set. $[\min_{Xi}, \max_{Xi}]$ is the distribution interval of random variables Xi in data set s, and $[new\_\min_{Xi}, new\_\max_{Xi}]$ is the distribution interval after normalization of random variables X_i. Usually, all variables X_i normalized in the interval of [0,1] to eliminate the influence of dimension.

(2) Standardization of Z-Score

$$x'_{li} = \frac{x_{li} - \mu_{Xi}}{\sigma_{Xi}} \tag{3.2}$$

where μ_{Xi} is the average value of random variables and σ_{Xi} is the standard deviation of random variables Xi.

3.2.2 Correction of Abnormal Values for Variables

In a single variable, the noise contained in manufacturing data is reflected in the deviation of the data value of the variable from the overall distribution of its variable, which is called outlier. These outliers will seriously affect the skewness of normalized data distribution. Especially, the max–min normalization method is particularly sensitive to outliers of variables, and the result of z-score normalization will also be affected by outliers. In this chapter, Rule 3.1 will be used to correct the abnormal values of variables.

Rule 3.1

$$If\ x_{li} > ub_{Xi},\ \ Then\ x_{li} = ub_{Xi}$$
$$If\ x_{li} < lb_{Xi},\ \ Then\ x_{li} = lb_{Xi}$$

In Rule 3.1, ub_{Xi} and lb_{Xi} are the upper and lower bounds of variable Xi, which are used to correct abnormal values of variables. Because the amount of historical data has reached a certain scale, it is impossible to use scatter plot method and hypothesis testing method suitable for small samples to detect abnormal values of variables. For ub_{Xi} and lb_{Xi}, this section introduces the 3σ method and quartile distribution method.

(1) 3σ method

From Chebyshev inequality, $\mathrm{P}(|V_i - \mu_{Xi}| \geq \varepsilon) \leq \sigma_{Xi}/\varepsilon^2, \varepsilon = 3\sigma_{Xi}\mathrm{P}(|X_i - \mu_{Xi}| \geq 3\sigma_{Xi}) \leq \sigma_{Xi}/9\sigma_{Xi}^2$, it can be known that $\mathrm{P}(|X_i - \mu_{Xi}| \geq 3\sigma_{Xi}) = 0.0027$, when Xi obeys normal distribution, it can be known that it Xi is distributed in the interval centered on the mean value $3\sigma_{Xi}$ with high probability. Therefore, set ub_{Xi} and lb_{Xi} as follows:

$$ub_{Xi} = \mu_{Xi+}3\sigma_{Xi} \tag{3.3}$$

$$ul_{Xi} = \mu_{Xi} - 3\sigma_{Xi} \tag{3.4}$$

(2) Quadrant distribution method

In outlier correction, the standard deviation is easily affected by outliers, so the quartile distribution method based on the distance between upper and lower quantiles is also a common method for outlier correction. $Q3_{Xi}$ is the upper quartile of a variable, $Q1_{Xi}$ is the lower quartile of a variable, and d_F is the distance between the upper and lower quartiles, which is called extreme difference. While ub Xi and lb_{Xi} can be set as follows:

$$d_F = Q3_{Xi} - Q1_{Xi} \tag{3.5}$$

$$ub_{Xi} = Q1_{Xi} - 1.5d_F \tag{3.6}$$

$$ul_{Xi} = Q3_{Xi} + 1.5d_F \tag{3.7}$$

3.3 Filling of Missing Data

3.3.1 Filling Method for Missing Data

The noise contained in manufacturing data is also manifested as data incompleteness, that is, the attribute values of many records are vacant. If the mth attribute of the ith record in the data set is a missing value, it is recorded as $x_{im} = \text{null}$. According to whether the records have missing values, the data sets can be divided into complete data sets and vacant data sets. According to whether there are missing values, variable sets can be divided into complete variable sets and vacant variable sets. Specific definitions are as follows:

$$S_{complete} = \{x_i \in \mathrm{S} | \forall j\,, x_{ij} \neq null, 1 \leq \mathrm{i} \leq M, 1 \leq \mathrm{j} \leq N\} \tag{3.8}$$

$$S_{miss} = \mathrm{S} - S_{complete} \tag{3.9}$$

$$X_{complete} = \{X_i \in \mathrm{X} | \forall l\,, x_{li} \neq null, 1 \leq l \leq M, 1 \leq i \leq N\} \tag{3.10}$$

$$X_{miss} = \mathrm{X} - X_{complete} \tag{3.11}$$

Although rough set and neural network have certain advantages in dealing with incomplete data sets, data-based modeling methods such as linear regression, decision tree and support vector machine can achieve more stable results on complete data sets. Therefore, it is necessary to design a robust missing value filling method for manufacturing data. The commonly used missing value filling techniques can be divided into the following categories:

(1) Rule-based filling method [25]

① Global constant filling method: for the variables X_i in X_{miss}, calculate the mean or median of the known data values to complete the missing values. This method will reduce the variance of variables when there are many missing values.

② Random number filling method: for the variables X_i in X_{miss}, the distribution is inferred from their known data values, and the missing values of variables are filled by random sampling according to the distribution. This method will increase the variance of variables when there are many missing values.

③ Delete variable filling method: delete the attributes X_{miss} corresponding to Chinese variables and keep the corresponding attributes $X_{complete}$. This way will lead to certain data loss.

④ Delete record filling method: delete the data records in and keep them. This way will lead to certain data loss.

⑤ Hot deck filling method: specify one or more sorting attributes, sort these attributes, find the data records matching the sorting attributes for the data records containing missing values, and fill the missing values in the vacancy attributes with the matched data records. If there are no data records on the match, reduce or re-specify the sorting properties in turn. This method requires more manual intervention.

(2) Filling method based on model

In the prediction-based filling method, a prediction model $X_{miss,i} = f_{imputate}(X_{complete})$ is constructed to predict the value of $X_{miss,i}$ in S_{miss} by training and parameter estimation, taking $S_{complete}$ as a training set, $X_{complete}$ as attribute variable and $X_{miss,i} \in X_{miss}$ as prediction variable. According to the difference of $f_{imputate}$, the model-based filling methods are as follows:

① Naive Bayes filling method: Naive Bayes classification model can fill discrete variables. The maximum likelihood method is used to estimate the model parameters, and the model construction speed is fast. It requires that the variables meet the following requirements: the variables are independent of each other and the distribution of the variables $X_{complete}$ is known.

② Decision tree filling method [26]: C4.5 decision tree can fill discrete variables $S_{complete}$. Firstly, the variable is discretized, the root node is selected according to the information gain of the variable, and the decision tree is constructed recursively, which makes the model construction faster. In order to avoid over-fitting, pruning techniques are usually used to prune decision trees.

③ Linear regression filling method [27]: Linear regression can fill continuous variables. The least square method is used to estimate the model parameters, and the model construction speed is fast, but the filled and neutral variables $X_{miss,i}$ and $X_{complete}$ have high linear correlation.

④ Neural network filling method [28]: Neural network can fill discrete and continuous variables. The training speed of model is slow when the network is trained by back propagation method. On the premise of optimizing the structure and parameters of the model, the nonlinear relationship between $X_{complete}$ and $X_{miss,i}$ and can be fitted, but it is easy to over-fit $S_{complete}$, which leads to inaccurate filling S_{miss}.

⑤ Support vector regression filling method [29]: Support vector regression filling method can be used to fill continuous variables, and a nonlinear support vector regression model is constructed using complete data sets to predict missing values. The support vector regression model trains the model by the method of sequence minimum optimization, and its training speed is faster than that of neural network. Support vector regression is an effective filling method.

(3) Distance-based filling method

KNN filling method [30]: KNN method is a commonly used inert learning method. For the data x_i records in S_{miss}, find K complete data records most similar to x_i from $S_{complete}$ by distance formula; record these K pieces of data in the vacancy attribute x_i filled by weighted average. In KNN filling method, the similarity measure between data records only considers variables of $X_{complete}$. KNN has the advantages of simplicity, no need of training and high precision, but every missing value of a data record needs to traverse the whole $S_{complete}$ and the filling speed is slow. Therefore, the filling method based on KNN is often used in combination with clustering method [31].

3.3.2 Memetic Algorithm and Memetic Calculation

When using intelligent algorithm to deal with data analysis problems, it is very time-consuming to calculate the fitness function of intelligent algorithm for dealing with data sets with large data scale and dimension. However, for some datasets, it is very difficult to optimize due to the influence of data distribution. Therefore, how to design intelligent algorithms with good performance and high efficiency is an important research issue in recent years. Meme is a concept defined by Dawkins in his book Selfish Gene [32], and its Chinese translations include "seeking mother" and "cultural gene". A meme represents a cultural evolution unit, and its carrier can be a book or a piece of music. Meme can improve individuals locally during their evolution. Meme and Darwin's theory of evolution are corresponding concepts, and biological evolution can be realized by gene recombination and mutation, while the spread of Meme can improve many individuals, thus forming cultural evolution. Human society is an evolutionary system combining biological evolution and cultural evolution.

When Moscato [33] studied the Travelling Sales Problem (TSP), it was found that genetic algorithm was difficult to design effective crossover and mutation operators to produce feasible solutions when solving TSP problems. As a result, the accuracy and efficiency of genetic algorithm in solving TSP problems are not high. For this reason, he introduced the concept of meme in the field of evolutionary computation for the first time. Meme represents the local improvement of individuals and is expressed as a local search operator in the algorithm. Experiments show that the combination of genetic algorithm and local search can effectively balance exploration and exploration of TSP problem and achieve better solution effect and efficiency. Although the effect on TSP is remarkable, the concept of Memetic algorithm has not been widely accepted since it was put forward but is regarded as an expression of hybrid intelligent algorithm. In the 1990s, researchers usually aim to design efficient algorithms. On the one hand, intelligent algorithms with swarm intelligence features such as Particle Swarm Optimization (PSO) [34] and Ant Colony Optimization (ACO) [35] have been put forward one after another, and many achievements have been accumulated in improving evolutionary algorithms and selecting parameters. In addition, many standard test sets are summarized for the well-known problems such as complex continuous function optimization and TSP, which are used as the benchmark to evaluate the performance of the algorithm. The research of intelligent algorithm design is realized by comparing the performance of different algorithms on these test sets.

The proposition of No Free Lunch Theorem (NFLT) [36] has changed the research pattern of evolutionary computation. All algorithms indicated by NFLT (including parameterized examples) have the same performance sum on all problems, which is expressed by Formula (3.12):

$$\sum_{f} P(x_m | f, A) = \sum_{f} P(x_m | f, B) \tag{3.12}$$

$P(x_m | f, A)$ represents the probability that algorithm A can find the optimal solution of optimization target f. $P(x_m | f, B)$ represents the probability that algorithm B can find the optimal solution of optimization target f. Equation (3.12) indicates that any pair of algorithms A and B have the same performance on all problems. Although NFLT makes it impossible to design the so-called "optimal" algorithm, it also reveals the existence of efficient algorithms for certain problems. Therefore, Memetic algorithm is regarded as an effective computational paradigm. The design of Memetic algorithm begins to develop into two types:

(1) The algorithm is designed for a specific class of problems, and then the Memetic algorithm for solving specific problems is proposed. Obviously, this kind of research method needs a deeper understanding of specific problems. For example, Memetic algorithm has been successfully applied to classical problems such as traveling salesman problem and flow shop scheduling problem [37].
(2) Develop and design robust algorithms that are suitable for most problems, with the Adaptive Coevolutionary Memetic Algorithm as a typical representative [38]. In the Adaptive Coevolutionary Memetic Algorithm, the relevant information of local search operators is encoded into Memes, and these Memes and populations coevolution together to adapt to the problem space during the operation of the algorithm to ensure the robustness of the algorithm. With the emergence of various Memetic algorithms, the theoretical research on Memetic algorithms is deepening day by day. Krasnogor defines Memetic algorithm as a special evolutionary algorithm, in which local search operators are used to improve individuals. The general framework of Memetic algorithm is shown in Fig. 3.2.

In this Memetic algorithm framework, besides the necessary initialization, basic elements such as optimization target evaluation and common operators of evolutionary computation such as recombination, mutation, and selection, it also includes local search for individuals. For example, Petalas implements Memetic PSO algorithm based on PSO and random walk local search.

Fig. 3.2 Pseudocode of memetic algorithm framework

procedure Memetic_algorithm()
begin
population initialization
localSearch
evaluation
do
    recombination
    mutation
    evaluation
    selection
while *the termination criterion is not satisfied*
end

According to the above definition and framework, Memetic algorithm has two characteristics:

(1) Memetic algorithm is firstly a population-based evolutionary algorithm.
(2) Meme means local search in Memetic algorithm. Ong [39] puts forward the concept of Memetic computing from the perspective of problem solving based on Memetic algorithm and coevolutionary algorithm and defines Memetic computing as a computational paradigm that uses meme to solve problems. Meme is defined as an information unit encoded in "calculation representation" in Memetic calculation, and exists in the form of operator, learning strategy, local search and so on. Memetic calculation is realized through the cooperation and interaction between memes. In Memetic calculation, the definition of meme is extended from local search to arbitrary search or learning strategy, and operators such as crossover and mutation can be called meme. In addition, Memetic computation does not require that the algorithm is based on population. Therefore, Memetic calculation is a generalization of Memetic algorithm.

According to Icca [40], the algorithms studied under the framework of Memetic computing are optimized by a variety of memes, which involve more operators and parameter settings, resulting in relatively complex algorithm design. Therefore, Icca introduced the idea of Ockham's Razor into Memetic computing, and Icca believes that the following three memes are sufficient and necessary for designing simple and efficient algorithms in Memetic computing:

- Long distance exploration: in the search process, a new solution is generated with a larger search step or a larger disturbance probability.
- Middle distance exploration: in the search process, a new solution is generated with moderate search step or moderate change probability.
- Short distance exploration: in the search process, a new solution is generated with a smaller search step or a smaller change probability.

In this chapter, when designing the intelligent algorithm for data analysis, we followed the design criteria of Icca, designed and selected the corresponding memes, and studied the cooperation between different memes under the framework of

Memetic algorithm and Memetic calculation. Two population-based Memetic algorithms are proposed: memetic PSO (GD-MPSO) based on Gaussian mutation and depth-first search, and memetic PSO (MCLPSO) based on extensive learning. The performance and efficiency of these two algorithms are verified by complex function optimization problems.

3.3.3 *Attribute Weighted K Nearest Neighbor Missing Value Filling Method (KNN) Based on Gaussian Mutation and Depth First Search (GD-MPSO): GD-MPSO-KNN*

K-nearest neighbor is a learning method of example learning or inert learning, which is widely used in missing value filling [41]. In this section, weighted KNN-based padding is adopted. To further improve the prediction accuracy of weighted KNN, feature weighting technology based on intelligent algorithm will be applied.

For the records $x_{im} = null$ of x_i, select the k closest records $neighbor_{i1}$, $neighbor_{i2}$, ..., $neighbor_{ik}$ from other records of the data set S, and these neighbor records are selected according to the similarity measure. In this section, the weighted Euler formula is used as a similarity measure, i.e., weighted k nearest neighbors (WKNN), which fw_j indicates the weight of the jth attribute of $X_{complete}$. The greater the value fw_j, the higher the weight of attribute j. The weighted sum of k neighbors of is obtained by Formula (3.13), $w_j = 1/d(x_i, neighbor_{ij})$, which $\hat{x}_{im}$ is the estimated value of x_{im}. For the convenience of discussion, this section assumes that only variables X_m contain vacant values, that is, $X_{miss} = \{X_m\}$, $X_{complete} = X - \{X_m\}$.

$$\hat{x}_{im} = \sum_{k=1}^{K} w_k neighbor_{ikm} / \sum_{k=1}^{K} w_k \tag{3.13}$$

$$\mathrm{d}(x_i, x_j) = \sqrt{\sum_{k=1}^{D} fw_j (x_{ik} - x_{jk})^2} \tag{3.14}$$

In this section, the filling method based on GD-MPSO and WKNN, GD-MPSO-WKNN, can be divided into two stages, namely, the first stage of training and the second stage of missing value filling.

The first stage: training stage, that is, using GD-MPSO to optimize the weight $\mathrm{fw_j}$ of each feature J to improve the prediction accuracy based on KNN method.

① Coding mode: the solution of particle i is coded into d-dimensional vector $solution_i$ [42], $solution_i = (fw_{i1}, fw_{i2}, \ldots, fw_{iD})$, $\mathrm{D} = |X_{complete}|$, which fw_{ij} is the weight assignment of the jth variable $X_{complete}$ *of* $solution_i$, $0 \le fw_{ij} \le 1$, which $solution_i$ is the weight assignment of all attributes. The position vector pos_i and optimal position $pbest_i$ of particle i can be expressed as $solution_i$.

② Objective function: GD-MPSO-KNN $S_{complete}$ is fitted by adjusting the weight of the medium variable $X_{complete}$ in the distance Formula (3.14). The objective function value of the solution $solution_i$ of particle i is determined by the Leave-One-Out cross-validation method. The specific solution steps are as follows:

Step 1. For each sample x_i in each sample $S_{complete}$, find the k nearest neighbors $neighbor_{i1}, neighbor_{i2}, \ldots, neighbor_{iK}$ from the sample S − $\{x_i\}$ by its weighted distance function in $X_{complete}$ (3.14). The value fw_j of the weight fw_{ij} in (3.14) is assigned as the jth component of $solution_i$.
Step 2. Take the weighted sum of $neighbor_{i1}, neighbor_{i2}, \ldots, neighbor_{iK}$ the values x_{im} on the mth attribute as the estimated value $\hat{x}_{im}$, that is (3.13).
Step 3. Calculate the estimated value of all records x on the mth attribute and take the minimum mean square error between the predicted value and the actual value as the objective function value of $solution_i$, that is, (3.15).

$$MSE\left(S_{complete}\right) = \sqrt{\frac{\sum_{X_i \in S_{complete}} \left(x_{im} - \hat{x}_{im}\right)^2}{\left|S_{complete}\right|}} \tag{3.15}$$

The objective function flow of GD-MPSO-KNN is shown in Fig. 3.3. Therefore, a set of D-dimensional feature weights $(w_1, w_2, \ldots, w_D)$ can be optimized by GD-MPSO-KNN.

In the second stage, the missing values are filled, that is, for each data record x_{im} in the data set S_{miss}, K nearest data records are found from the data set $S_{complete}$ according to Formula (3.14), and the estimation $\hat{x}_{im}$ of the missing values is obtained by Formula (3.13), thus completing the missing values filling in the data set. The implementation process of GD-MPSO-WKNN is shown in Fig. 3.4.

```
function f(solution: sol)
begin
   for each xi in Scomplete
      find K nearest instance in Scomplete-{xi} for xi by (3.13)
      /*the weight fwj for the jth variable in Xcomplete is specified by jth value of sol*/
      compute x̂im for xim by (3.12)
   endfor
   return the result of (3.14) as the value of f
end
```

Fig. 3.3 Optimization objective function f of GD-mpso-KNN

```
procedure GD-MPSO-WKNN()
begin
   call GD-MPSO to evolving a vector of weights (w_1, w_2, ..., w_D)
   for each x_i in S_miss
      find K nearest instance in S_complete for x_i by (3.13)
      /*the weight fw_j for the jth variable in X_complete is specified by w_j */
      compute x̂_im for x_im by (3.12)
   endfor
end
```

Fig. 3.4 Implementation of pseudocode by GD-MPSO-WKNN

3.3.4 Numerical Verification

To verify the filling accuracy of GD-MPSO-KNN, the sensor data set D2 with the most vacant values in the manufacturing system is used as the test set. Specific experimental verification steps are as follows:

Step 1: For three sensor attributes (X_5, X_{12}, X_{204}) with large Coefficient of Variation, CV), i.e., the ratio of standard deviation to mean value, the missing values are randomly marked according to the missing value ratio of 10, 20, 30, 40, 50%.
Step 2: Call GD-MPSO-WKNN or other methods to complete this set of marked missing values.
Step 3: Evaluate the filling accuracy according to the Mean Square Error (MSE) and Mean Absolutely Error (MAE) of the missing value estimator and the missing value original value.

$$MSE(S_{miss}) = \sqrt{\frac{\sum_{X_i \in S_{miss}} (x_{im} - \hat{x}_{im})^2}{|S_{miss}|}} \tag{3.16}$$

$$MAE(S_{miss}) = \frac{\sum_{X_i \in S_{miss}} |x_{im} - \hat{x}_{im}|}{|S_{miss}|} \tag{3.17}$$

To objectively evaluate the filling accuracy of GD-MPSO-WKNN, GD-MPSO-WKNN is compared with the following methods:

(1) Model-based filling methods: Linear Regression (LR) filling method, Support Vector Regression (SVR) filling method.
(2) Filling method based on distance: KNN filling method.

The maximum iteration number of GD-MPSO-KNN is set to 100, K = 20 of k nearest neighbor in optimization target f, and the parameter settings are shown in Table 3.2.

See Table 3.3 for the results of filling missing values. Analysis shows that:

Table 3.2 Algorithm parameter settings

Function				
$f1$	[0, 0, …, 0]	0	$[-100, 100]^D$	$[-100, 50]^D$
$f2$	[0, 0, …, 0]	0	$[-100, 100]^D$	$[-100, 50]^D$
$f3$	[0, 0, …, 0]	0	$[-100, 100]^D$	$[-100, 50]^D$
$f4$	[0, 0, …, 0]	0	$[-10, 10]^D$	$[-10, 5]^D$
$f5$	[0, 0, …, 0]	0	$[-100, 100]^D$	$[-100, 50]^D$
$f6$	[1, 1, …, 1]	0	$[-2.048, 2.048]^D$	$[-2.048, 1]^D$
$f7$	[0, 0, …, 0]	0	$[-32.768,32.768]^D$	$[-32.768,16]^D$
$f8$	[0, 0, …, 0]	0	$[-600, 600]^D$	$[-600, 300]^D$
$f9$	[0, 0, …, 0]	0	$[-100, 100]^D$	$[-100, 50]^D$
$f10$	[0, 0, …, 0]	0	$[-0.5, 0.5]^D$	$[-0.5, 0.2]^D$
$f11$	[0, 0, …, 0]	0	$[-100, 100]^D$	$[-100, 50]^D$
$f12$	[0, 0, …, 0]	0	$[-5.12, 5.12]^D$	$[-5.12, 2]^D$
$f13$	[0, 0, …, 0]	0	$[-5.12, 5.12]^D$	$[-5.12, 2]^D$
$f14$	[420.96, 420.96, …, 420.96]	0	$[-500, 500]^D$	$[-500, 500]^D$
$f15$	[1, 1, …1]	0	$[-50, 50]^D$	$[-50, 50]^D$
$f16$	[1, 1, …1]	0	$[-50, 50]^D$	$[-50, 50]^D$

Table 3.3 Results of missing value filling for X_5

MSE	10%	20%	30%	40%	50%
LR	1.46E+01	3.19E+01	1.22E+02	7.76E+01	5.64E+01
SVR	1.24E+01	2.23E+01	8.48E+00	6.14E+00	4.10E+01
KNN	1.01E+01	9.49E+00	8.63E+00	7.68E+00	8.01E+00
GS-MPSO-WKNN	**8.98E+00**	**9.15E+00**	**8.16E+00**	**6.75E+00**	**7.78E+00**
MAE	10%	20%	30%	40%	50%
LR	7.93E+00	9.66E+00	6.03E+00	1.12E+01	9.89E+00
SVR	**7.16E+00**	8.17E+00	**5.49E+00**	6.19E+00	8.41E+00
KNN	8.14E+00	7.73E+00	6.85E+00	5.97E+00	6.18E+00
GS-MPSO-WKNN	7.18E+00	**7.38E+00**	6.49E+00	**5.17E+00**	**5.95E+00**

- When the missing value ratio is 10%, the SVR filling method has the highest accuracy, but when the missing value ratio increases, the degradation of the SVR filling method is very obvious. With the increase of the missing value ratio and the reduction of learning samples, the SVR prediction model will fall into over-fitting;
- The change of LR filling accuracy is similar to that of SVR, but the accuracy of LR filling method is inferior to that of SVR filling method in every missing

Table 3.4 Results of missing value filling for X_{12}

MSE	10%	20%	30%	40%	50%
LR	4.47E+00	9.44E+01	8.61E+01	7.04E+01	9.77E+01
SVR	**3.20E+00**	1.22E+01	2.03E+01	1.70E+01	8.12E+01
KNN	3.39E+00	2.88E+00	2.68E+00	2.52E+00	2.55E+00
GS-MPSO-WKNN	3.24E+00	**2.73E+00**	**2.52E+00**	**2.37E+00**	**2.40E+00**
MAE	10%	20%	30%	40%	50%
LR	2.98E+00	1.33E+01	1.28E+01	9.34E+00	1.05E+01
SVR	**2.38E+00**	3.66E+00	4.67E+00	3.39E+00	9.21E+00
KNN	2.69E+00	2.26E+00	2.14E+00	1.98E+00	2.02E+00
GS-MPSO-WKNN	2.54E+00	**2.15E+00**	**2.01E+00**	**1.85E+00**	**1.88E+00**

rate. Obviously, simple linear model is not suitable for complex sensor data filling problem.

- Compared with SVR filling method, KNN filling method has lower accuracy when the missing value ratio is small, but with the increase of missing value ratio, KNN filling method shows better robustness, and can achieve stable filling accuracy when the missing value ratio reaches 20, 30, 40, and 50%;
- Compared with KNN, GD-MPSO-WKNN has higher accuracy in every missing value ratio. When the missing value ratio is 10%, the filling accuracy of GD-MPSO-WKNN is close to SVR filling method. With the increase of the proportion of missing values, GD-MPSO-WKNN, like KNN, maintains high robustness and achieves high filling accuracy. GD-MPSO-WKNN uses the same decision-making method as KNN, which can effectively avoid overfitting. At the same time, GD-MPSO-WKNN makes full use of complete data to extract attribute weights and gives higher weights to attributes that significantly affect missing values. Therefore, GD-MPSO-WKNN is very suitable for filling the missing values of sensors in manufacturing system (Tables 3.4 and 3.5).

3.4 Outlier Detection Based on Data Clustering Analysis

3.4.1 Outlier Detection Based on Data Clustering

Data clustering or clustering analysis is to assign data records to different clusters. Data records belonging to the same cluster have higher similarity, while data records belonging to different clusters have lower similarity. Clustering is an important task of exploratory data mining, which has a high application in many fields of data analysis. Clustering data sets S into K classes can be defined as dividing datasets S into K blocks to represent the clustering results $Partition_K$ of datasets, namely:

Table 3.5 Results of missing value filling for X_{204}

MSE	10%	20%	30%	40%	50%
LR	1.15E+02	3.27E+02	4.53E+02	5.58E+02	6.89E+02
SVR	1.13E+02	2.96E+02	2.74E+02	5.04E+02	6.65E+02
KNN	1.14E+02	8.71E+01	7.50E+01	2.52E+01	2.55E+01
GS-MPSO-WKNN	**1.12E+02**	**8.67E+01**	**7.23E+01**	**2.37E+01**	**2.40E+01**
MAE	10%	20%	30%	40%	50%
LR	4.07E+01	6.62E+01	7.10E+01	7.47E+01	8.56E+01
SVR	**3.85E+01**	6.00E+01	5.01E+01	6.29E+01	7.56E+01
KNN	4.66E+01	4.18E+01	3.81E+01	3.63E+01	3.53E+01
GS-MPSO-WKNN	4.26E+01	**3.94E+01**	**3.29E+01**	**3.18E+01**	**3.02E+01**

$$Partition_K = (Cluster_1, Cluster_2, \ldots, Cluster_K), \forall k, Cluster_k \subseteq S, 1 \le k \le K$$

And satisfy the following constraints:

$$\forall k, Cluster_k \ne \varnothing, 1 \le k \le K$$

$$Cluster_i \cap Cluster_j = \varnothing, 1 \le i, j \le K, i \ne j$$

$$\bigcup_{k=1}^{K} Cluster_k = S$$

Clustering defined above is usually called hard clustering. In hard clustering, each data record belongs to a specific cluster. Clustering is realized by clustering algorithm according to similarity measure, and clustering results are evaluated by clustering criteria. Generally speaking, clustering criteria are also defined according to similarity measure. In this section, Euclidean distance formula is used as similarity measure, see Formula (3.18), and Formula (3.19) based on Euclidean distance is used as clustering criterion, where $centroid_j$ is the clustering center of $Cluster_j$. The smaller the value of $J(Partition_K)$, the higher the cohesion of each cluster.

$$\mathrm{d}(x_i, x_j) = \sqrt{\sum_{k=1}^{D} (x_{ik} - x_{jk})^2} \tag{3.18}$$

$$J(Partition_K) = \frac{\sum_{Cluster_j \in Partition_K} [\sum_{x_i \in Cluster_j} d(x_i, centroid_j)] / |Cluster_j|}{K} \tag{3.19}$$

Outlier detection based on clustering can be realized by Rule 3.2. When the distance between a data sample and the cluster center exceeds a certain threshold, the sample is considered as outlier. Distance threshold α can be defined as 3 times average clustering center, as shown in Formula (3.20).

Rule 3.2

$$If\ x_i \in Cluster_k \wedge d(x_i, centroid_k) > \alpha,\ Then\ x_i\ is\ outlier$$

$$\alpha = 3 * \left[\sum_{x_i \in Cluster_j} d(x_i, centroid_j) \right] \Big/ |Cluster_j| \tag{3.20}$$

3.4.2 K-Means Clustering

K-MEANS clustering is one of the simplest and most commonly used clustering algorithms [43], in which the number of clusters k is specified by the user. The specific steps are as follows:

Step 1: For cluster, $Cluster_1$, $Cluster_2$, ..., $Cluster_K$, initial cluster center, $(centroid_1, centroid_2, centroid_K,)$;
Step 2: Assign each data x_i record in S to a cluster $Cluster_j$ that $d\left(x_i, centroid_j\right)$ can be minimized.
Step 3: For each cluster, use Formula (3.21) to update its cluster center.

$$centroid_j = \frac{1}{\left|Cluster_j\right|} \sum_{x_i \in Cluster_j} x_i \tag{3.21}$$

Step 4: Repeat Step 3 and Step 4 until the cluster centers of all clusters do not change.

It can be seen from the above steps that K-means clustering is a process of iteratively allocating data records and updating cluster centers. See Fig. 3.5 for specific implementation details of k-means.

Because K-means algorithm is simple and fast, it is widely used, but it has the following shortcomings:

① Improper selection of clustering number k will lead to poor clustering results;
② The clustering result depends on the selected similarity measure;
③ The clustering results are sensitive to the initial k clustering centers.

K is generally adjusted by trial and error, and the selection of similarity measure depends on the prior knowledge of data set. This section discusses the third problem, that is, K-means algorithm is sensitive to initial clustering, which makes clustering

```
procedure KMEANS(K initial centroids: centroid_1, centroid_2, ..., centroid_K)
  do
    for i=1 to K
        Cluster_i = ∅
    endfor
    for i=1 to M
        find the cluster Cluster_j that minimize d(x_i, centroid_j)
        Cluster_j = {x_i} ∪ Cluster_j
    endfor
    for i=1 to K
        recalculate the centroid_j according to (3.19)
    endfor
  while (the stop criterion is not met)
end
```

Fig. 3.5 Implementation of pseudocode by K-means algorithm

criterion function fall into local optimum, while optimization method can be used to set initial clustering center, weaken the sensitivity of algorithm to initial clustering center and further minimize clustering criterion function.

3.4.3 Data Clustering Algorithm Based on GS-MPSO and K-Means Clustering (GS-MPSO-KMEANS)

GD-MPSO uses depth-first search, which is inefficient in high-dimensional problem optimization. Therefore, GD-MPSO's depth-first search Deepest_local_search is replaced by local search SA_local_search based on simulated annealing in memorable comprehensive learning PSO (mclpso), which is called GS-MPSO. GS-MPSO uses the following search methods:

① Long-distance detection: PSO with compression factor;
② Medium distance detection: Gaussian mutation operator;
③ Short distance detection: local search based on simulated annealing.

GS-MPSO uses the same meme cooperative interaction strategy as GD-MPSO. In each generation of PSO evolution, SA_local_search is only applied to the hopeful particles, and the promising areas are searched with fine granularity. However, mutation operator is only applied to stagnant particles. Because stagnant particles can't improve their pbest i from their neighbors, it makes stagnant particles jump and search for new areas.

GS-MPSO-KMEANS is a clustering algorithm based on GS-mpso and KMEANS, which minimizes the clustering criterion function by optimizing the initial clustering center of KMEANS.

```
function f(solution: sol)
begin
    decompose sol and get centroid_1, centroid_2, ..., centroid_K
    Partition_K=KMEANS(centroid_1, centroid_2, ..., centroid_K)
    return the result of J(Partition_K) as the value of f
end
```

Fig. 3.6 Realization of pseudocode by objective function F of MCLPSO-KME ANSVAR

(1) Coding mode: the solution of particle I is coded into d-dimensional vector, $D = K * N$, where k is the number of clusters and n is the data dimension [44]. $solution_i = (centroid_{i1}, centroid_{i2}, \ldots, centroid_{iK})$, $centroid_{iK}$ is the initial assignment of the solution $solution_i$ of particle i to the cluster center $centroid_k$ of the kth cluster, and the solution of particle i gives the initial value of the cluster center of each cluster. The position vector pos_i and optimal position of particle i $pbest_i$ can be expressed as$solution_i$:
(2) Target function: GS-MPSO-KMEANS optimizes the clustering criteria $J(Partition_K)$ by adjusting the initial clustering center of KMEANS, $centroid_{i1}, centroid_{i2}, \ldots, centroid_{iK}$ to improve the quality of variable clustering. It is easy to decompose the solution of particle i into k clustering centers, $centroid_{i1}, centroid_{i2}, \ldots, centroid_{iK}$, and call KMEANS to get variable clustering $Partition_K$ and its clustering criteria $J(Partition_K)$ as parameters, and take the $J(Partition_K)$ as objective function values.

According to the above discussion, the objective function flow of GS-MPSO-KMEANS is given in Fig. 3.6.

3.4.4 Numerical Verification

To verify the clustering performance of GS-MPSO-KMEANS, this section uses D_1 and D_2 datasets for verification. The number of clusters is set to 5, 10 and 15 respectively. Choose KMEANS and compare cf-PSO-KMEANS with GS-MPSO-KMEANS based on cf-PSO and kmeans.

The maximum iteration number of GS-MPSO-KMEANS is set to 100, and other parameter settings are consistent with Table 3.2. For each dataset, each algorithm is run 100 times. Table 3.6 shows the mean and variance of the optimized value of clustering criterion function by each algorithm.

According to Table 3.6, KMEANS without optimized initial clustering center has a big gap compared with cf-PSO-KMEANS and GS-MPSO-KMEANS, two other intelligent algorithms for optimized initial clustering center. When the number of clusters increases, both GS-MPSO-KMEANS and cf-PSO-KMEANS can find more compact clusters to further optimize the clustering criteria, but KMEANS cannot

Table 3.6 Data clustering results

	D_1 ($K = 5$)	D_1 ($K = 10$)	D_1 ($K = 20$)	D_2 ($K = 5$)	D_2 ($K = 10$)	D_2 ($K = 20$)
KMEANS	1.18E+00 ± 3.70E−02	1.05E+00 ± 3.10E−02	1.01E+00 ± 6.02E−02	2.32E+00 ± 2.69E−01	2.37E+00 ± 1.95E−01	2.19E+00 ± 7.36E−01
GS-MPSO	**9.27E−01 ± 5.04E−02**	**7.54E−01 ± 6.33E−02**	**7.22E−01 ± 3.18E−02**	**6.17E−01 ± 6.04E−02**	**5.03E−01 ± 5.34E−02**	**4.48E−01 ± 8.50E−02**
cf-PSO	9.34E−01 ± 5.71E−02	8.28E−01 ± 3.80E−02	7.34E−01 ± 4.36E−02	6.53E−01 ± 1.19E−01	5.34E−01 ± 7.12E−01	4.92E−01 ± 7.12E−01

further optimize the clustering criteria when the number of clusters increases. GS-MPSO-KMEANS has stronger ability to optimize clustering criteria than cf-PSO-KMEANS, but when D_1 (K = 5), GS-MPSO-KMEANS has no obvious improvement compared with cf-PSO-KMEANS, which is due to the small number of samples in D_1, and when the number of clusters is small, there are relatively few possible cluster combinations, cf-PSO-KMEANS. However, in the cases of D_1 (K = 10), D_2 (K = 5), D_2 (K = 10), D_2 (K = 20), the optimization ability of GS-MPSO-KMEANS is significantly improved compared with that of cf-PSO-KMEANS, and it can effectively reduce the variance while improving the average clustering criterion function, which shows that GS-MPSO-KMEANS can effectively reduce the variance.

3.5 Redundant Variable Detection Based on Variable Clustering

3.5.1 Principal Component Analysis

In variable clustering, the first principal component of a group of variables is usually obtained by principal component analysis (PCA) [45] as the clustering center of this group of variables. PCA is a common dimension reduction method, which can use fewer attributes instead of more attributes, and these fewer attributes can reflect the information of more attributes as much as possible, and they are linearly independent of each other. In essence, PCA is realized by coordinate transformation, and the variance can be maximized along the new coordinate axes, and the projection of data on these coordinate axes is the principal component. The basic principle of PCA is as follows:

Given the data set, PCA transforms the samples into (3.22).

$$S = \{x_i\}_{i=1}^{M},\ x_i \in R^N x_i x_i'$$

$$x_i' = x_i U' \tag{3.22}$$

In which, the subarrays U' of partial columns are selected from U, U is an $N \times N$ dimensional orthogonal matrix, and the jth column U_j is the jth feature vector of the sample covariance matrix C.

C is the sample covariance matrix of data set s, which is defined in Formula (3.23).

$$C = \left(c_{ij}\right)_{N\times N} c_{ij}$$

$$c_{ij} = \frac{1}{M-1}\sum_{k=1}^{M}\left(x_{ki} - \mu_{Xi}\right)\left(x_{kj} - \mu_{Xj}\right) \tag{3.23}$$

According to Formula (3.23), C is a real symmetric matrix, and according to the properties of the real symmetric matrix, c has n real characteristic roots (including multiple roots), and the characteristic vectors are all real vectors, which are different and corresponding. Equation (3.24) can be obtained by eigenvalue decomposition.

$$\lambda_j U_j = C U_j, \quad j = 1, 2, \ldots, N, \quad \lambda_1 \geq \lambda_2 \geq \cdots \geq \lambda_N \tag{3.24}$$

where λ_j is the eigenvalue of C and the corresponding eigenvector U_j. When $\lambda_1 \geq \lambda_2 \geq \cdots \geq \lambda_N$, order U $= (U_1, U_2, \ldots, U_N)$ according to the corresponding eigenvalue from big to small. The projection of S in the direction U_1 has the largest variance, the projection of S in the direction U_2 has the second largest variance, and so on. The eigenvectors are orthogonal in pairs, and T eigenvectors are selected from u according to the corresponding eigenvalues, $U' = (U_1, U_2, \ldots, U_T)$, and the data set after dimension reduction is the projection of U'.

$$S' = SU' \tag{3.25}$$

According to the above principles, it is easy to sum up the solving steps of first principal component:

Step 1: Obtaining a sample covariance matrix C of a sample dataset S;
Step 2: Calculate the eigenvalue of C according to Jacobi iterative method; $\lambda_1, \geq \lambda_2 \geq \cdots \geq \lambda_N$;
Step 3: Select the eigenvector U_1 corresponding to the maximum eigenvalue λ_1;
Step 4: Calculate the projection of the sample data in direction U_1 to get the First Principal Component (FPC): FPC$(S) = SU_1$.

Based on the above discussion, the pseudo code of first principal component is shown in Fig. 3.7.

```
function FPC(Dataset: S)
begin
   compute covariance matric C for S
   compute eigenvalues λ1, λ2, ..., λN for C, and λ1 ≥ λ2 ≥ ··· ≥ λN
   get eigenvectors U1, U2, ..., UN
   return the SU1 as the first principle component of S
end
```

Fig. 3.7 Pseudo-code of first principal component analysis

3.5.2 Variable Clustering Based on K-Means Clustering and PCA

Variable clustering is a method of exploratory data mining, which realizes redundant variable detection by clustering variables with strong linear correlation. Clustering variables into k classes can be defined as dividing the variable set x into k blocks. In this section, the clustering results of the variable set are expressed by $Partition_K$.

$$Partition_K = (Cluster_1, Cluster_2, \ldots, Cluster_K), \forall k, Cluster_K \subseteq X, 1 \le k \le K$$

And satisfy the following constraints:

$$\forall k, Cluster_k \ne \emptyset, \quad 1 \le k \le K$$
$$Cluster_i \cap Cluster_j = \emptyset, \quad 1 \le i, \quad j \le K, \ i \ne j$$
$$\bigcup_{k=1}^{K} Cluster_k = X$$

The variables belonging to the same cluster have strong correlation, while the variables belonging to different clusters have weak correlation. The most famous variable clustering tool is VARCLUS process in SAS software [46]. This section mainly introduces KMEANSVAR, a variable clustering algorithm based on k-means clustering. KMEANSVAR defines the distance measure between variables, the update of cluster center and the clustering criterion as follows:

(1) Distance: the distance d(.) (3.27) between variables is defined by Pearson correlation coefficient (3.26). The higher the correlation (whether positive correlation or negative correlation); The closer the distance between variables is; The smaller the correlation, the farther the distance between variables.

$$Pearson(X_i, X_j) = \frac{\sum_{k=1}^{M}(x_{ki} - \mu_{Xi})(x_{kj} - \mu_{Xj})}{\sqrt{\sum_{k=1}^{M}(x_{ki} - \mu_{Xi})^2}\sqrt{\sum_{k=1}^{M}(x_{kj} - \mu_{Xj})^2}} \quad (3.26)$$

$$d(X_i, X_j) = 1 - Pearson(X_i, X_j)^2 \quad (3.27)$$

(2) Updating the cluster center: obtaining the first principal component of variables belonging to the same cluster and updating the cluster center.

$$Cluster_k = \{X_{k1}, X_{k2}, \ldots, X_{kP}\}$$

Datasets $S_{Cluster\_k}$ are sample data composed of m observations of random vectors composed of cluster member variables $(X_{k1}, X_{k2}, \ldots, X_{kP})$. That is, the data set obtained by retaining the corresponding attributes $X_{k1}, X_{k2}, \ldots, X_{kP}$ on the original data set s and deleting other attributes. According to the method of finding first principal component in the previous section, $FPC(S_{Cluster_k})$ will be used as a new clustering center of $Cluster_k$.

(3) Clustering criteria: Through the above definition of variable similarity and updating rules of variable clustering centers, clustering criteria can be defined to measure the quality of variable clustering, and good clustering will maximize the clustering criteria. The homogeneity $H(Cluster_k)$ of a cluster $Cluster_k$ can be measured by the sum of squares of Pearson correlation between the member variables of the cluster and the cluster center $centroid_k$ of $Cluster_k$, as shown in Formula (3.28), where is the first principal component of the cluster center of the cluster.

$$H(Cluster_k) = \sum_{X_{ki} \in Cluster_k} Pearson^2(X_{ki}, centroid_k) \tag{3.28}$$

Cluster homogeneity: obtained by summing the homogeneity indexes of all clusters $H(Partition_k)$ formed by this cluster (3.29).

$$H(Partition_k) = \sum_{Cluster_k \in Partition_k} H(Cluster_k) \tag{3.29}$$

Based on the above discussion, KMEANSVAR can be implemented as follows:

Step 1: Initialize the cluster centers $centroid_1, centroid_2, \ldots, centroid_K$ of K clusters $Cluster_1, Cluster_2, \ldots, Cluster_K$;
Step 2: Empty each cluster;
Step 3: For each variable X_i in the variable set X, find the X_i's nearest cluster $Cluster_{nearest}$ through (3.30) and assign it to $Cluster_{nearest}$;

$$Cluster_{nearest} = \arg \min_{Cluster_K \in Partition_k} (d(X_i, Cluster_k)) \tag{3.30}$$

$$Cluster_{nearest} = Cluster_{nearest} \cup X_i \tag{3.31}$$

Among them, the distance between variable and cluster is defined as the distance between variable and cluster center, namely:

$$\mathrm{d}(X_i, Cluster_k) = \mathrm{d}(X_i, centroid_k)$$

Step 4: For clustering clusters $Cluster_k$, the first principal component of its member variables $FPC(S_{Cluster\_k})$ is obtained as the new clustering center $centroid_k$ of clustering clusters $Cluster_k$;
Step 5: Iterate Step 2 to Step 4 repeatedly until the cluster center of each cluster does not change or reaches the maximum number of iterations.

```
procedure KMEANSVAR(K initial centroids: centroid_1, centroid_2, ..., centroid_K)
  do
    for k=1 to K
      Cluster_k = Ø
    endfor
    for i=1 to N
      Cluster_nearest = argmin_{ClusterkεPartitionK}(d(X_i, Cluster_k))
      Cluster_nearest = {x_i} ∪ Cluster_nearest
    endfor
    for k=1 to K
      recalculate the centroid_k by FPC(S_{Cluster_k})
    endfor
  while (the stop criterion is not met)
  return Partition_K = {Cluster_1, Cluster_2, ..., Cluster_k} as result
end
```

Fig. 3.8 KMENASVAR algorithm

Based on the above discussion, the pseudo code of KMEANSVAR is shown in Fig. 3.8.

3.5.3 *Variable Clustering Algorithm Based on MCLPSO (MCLPSO-KMEANSVAR)*

Although variable clustering algorithm based on K-means clustering and PCA can cluster variables effectively, like traditional K-means algorithm, KMEANSVAR is sensitive to the initial clustering center and easily falls into local optimum, resulting in poor clustering quality. To overcome this deficiency, this section proposes a variable clustering algorithm based on MCLPSO, MCLPSO-KMEANSVAR,

(1) Coding mode: the solution of particle i is coded into d-dimensional vector, where D = K*M, k is the number of clusters and m is the number of observed values of variables. $centroid_{iK}$ is the initial assignment of the solution $solution_i$ of particle i to the cluster center of the kth cluster $centroid_k$, and the solution of particle i gives the initial value of the cluster center of each cluster. The position vector pos_i and optimal position of particle i $pbest_i$ can be expressed as $solution_i$.

$$solution_i = (centroid_{i1}, centroid_{i2}, \ldots, centroid_{iK})$$

(2) Objective function: MCLPSO-KMEANSVAR optimizes the clustering criteria $H(Partition_K)$ by adjusting the initial clustering center of KMEANSVAR to improve the quality of variable clustering. It is easy to decompose the solution of

function *f(solution: sol)*
begin
decompose sol and get $centroid_1, centroid_2, \ldots, centroid_K$
$Partition_K = KMEANSVAR(centroid_1, centroid_2, \ldots, centroid_K)$
return the result of $1/H(Partition_K)$ *as the value of f*
end

Fig. 3.9 Objective function of MCLPSO-KMEANSVAR

particle i into k cluster centers, $centroid_{i1}, centroid_{i2}, \ldots, centroid_{iK}$, and call KMEANSVAR to get variable clustering $Partition_K$ and its clustering criteria $H(Partition_K)$ as the objective function value$1/H(Partition_K)$.

Based on the above discussion, the objective function flow of MCLPSO-KMEANSVAR is given in Fig. 3.9.

3.5.4 Numerical Verification

To verify the clustering performance of MCLPSO-KMEANSVAR, this section uses D1 and D2 data sets for verification. The number of clusters is set to 5, 10 and 15 respectively. Choose KMEANSVAR, and compare CLPSO-KMEANS with mCLPSO-KMEANSvar, a data clustering algorithm based on clpso and kmeans.

The maximum number of iterations of MCLPSO-KMEANSVAR is set to 100, so Chaotic_local_search will not be called in MCLPSO-KMEANSVAR. Other parameter settings are consistent with Table 3.2. For each data set, each algorithm is run 100 times. The mean and variance of optimized values of clustering criteria by each algorithm are shown in Table 3.7.

It can be seen from Table 3.7 that KMEANSVAR without optimized initial clustering center is far behind the other two intelligent algorithms CLPSO-KMEANSVAR and MCLPSO-KMEANSVAR in optimizing the initial clustering center when performing variable clustering on a large number of high-dimensional and practical manufacturing system data sets D_1 and D_2. MCLPSO-KMEANSVAR has stronger ability to optimize clustering criteria than CLPSO-KMEANSVAR, but on D_1 and D_2, MCLPSO-KMEANSVAR has almost no advantage when the number of clusters is 5, because the fewer the number of clusters, the fewer possible cluster combinations, so it is easy to search for better clusters through intelligence, but KMEANS is not ideal for optimizing clustering criteria function even when the number of clusters is small. When the number of clusters increases, the optimization ability of MCLPSO-KMEANSVAR is reflected. MCLPSO-KMEANSVAR can not effectively reduce the variance while optimizing the clustering criteria. According to the boxplot distribution of MCLPSO-KMEANSVAR, CLPSO-KMEANSVAR and KMEANSVAR in D_2 clustering, KMEANSVAR is the least stable. The result distribution of CLPSO-KMEANSVAR tends to be flat, and its performance is more

Table 3.7 Variable clustering results

	D_1 ($K = 5$)	D_1 ($K = 10$)	D_1 ($K = 20$)	D_2 ($K = 5$)	D_2 ($K = 10$)	D_2 ($K = 20$)
MCLPSO-KMEANSVAR	**3.44E+01 ± 2.12E−01**	**4.45E+01 ± 4.06E−01**	**5.15E+01 ± 5.29E−01**	**4.36E+01 ± 4.01E−01**	**6.51E+01 ± 4.22E+00**	**1.02E+02 ± 3.04E+00**
CLPSO-KMEANSVAR	**3.44E+01 ± 2.10E−01**	4.44E+01 ± 3.92E−01	5.13E+01 ± 5.12E−01	4.35E+01 ± 6.01E−01	6.41E+01 ± 3.42E+00	1.01E+02 ± 3.04E+00
KMEANSVAR	3.41E+01 ± 4.40E−01	4.32E+01 ± 1.43E+00	4.44E+01 ± 1.99E+00	3.99E+01 ± 2.64E+00	6.00E+01 ± 4.64E+00	4.94E+01 ± 3.44E+00

stable. However, when the clustering problem is complex (K = 10, K = 20), it is not difficult to find from Figs. 3.10 and 3.12 that the optimization result of MCLPSO-KMEANSVAR is better than that of CLPSO-KMEANSVAR, and MCLPSO-KMEANSVAR has a higher probability (Fig. 3.11).

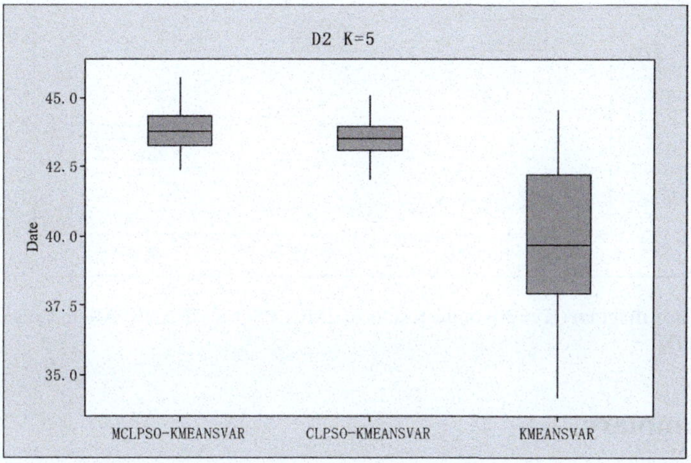

Fig. 3.10 Box diagram (K = 5) of the results of MC LPSO-KMEANSVAR algorithm running on D_4

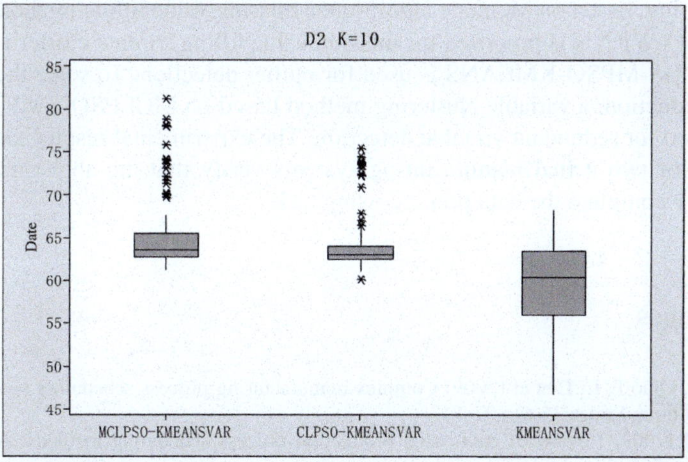

Fig. 3.11 Box diagram (K = 10) of the results of 1 MC LPSO-KMEANSVAR algorithm running on D_4

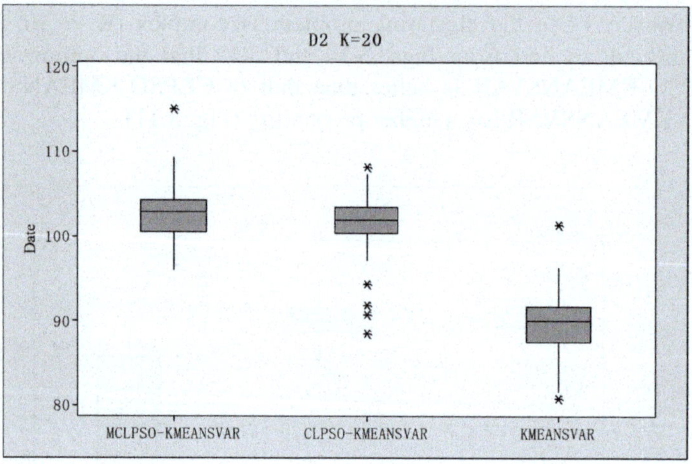

Fig. 3.12 Box diagram (K = 20) of the results of 2MCLPSO-KMEANSVAR and other algorithms running on D_4

3.6 Summary

This chapter focuses on the data preprocessing technology for data standardization, data cleaning, data specification and other issues. Firstly, the common data normalization methods based on rules are introduced. Aiming at the problem of data cleaning, based on Memetic algorithm, a missing value filling method based on GD-MPSO-WKNN is proposed for missing value filling. A data clustering method based on GS-MPSO-KMEANS is used for outlier detection. To solve the problem of data reduction, a variable clustering method based on MCLPSO-KMEANSVAR is proposed for redundant variable detection. The experimental results based on the data sets of two actual manufacturing systems verify that the above method can effectively complete the data preprocessing task.

References

1. Wu Q, Qiao F, Li L et al (2009) Complex manufacturing process scheduling based on data. Acta Autom Sinica 35(6):807–813
2. Liu M (2009) Review of data-based production process scheduling methods. Acta Autom Sinica 6:785–806
3. Chai T (2009) Challenges of the whole manufacturing process optimization control to the control and optimization theory and method. Acta Autom Sinica 6:641–649
4. Liu Q, Chai T, Qin S et al (2010) Overview of industrial process monitoring and fault diagnosis based on data and knowledge. Control Decis 25(6):801–807
5. Wang H, Qi C, Wei Y et al (2009) A summary of data-based decision-making methods. Acta Autom Sinica 35(6):820–833

6. Gao C, Yan L, Chen J et al (2009) Data-driven modeling and prediction algorithm for complex blast furnace ironmaking process. Acta Autom Sinica 35(6):725–730
7. Liu Y, Zhao J, Wang W et al (2009) Application of improved echo state network based on data in prediction of blast furnace gas production. Acta Autom Sinica 35(6):731–738
8. Gui W, Yang C, Li Y et al (2009) Optimization and application of data-driven copper flash smelting process operation mode. Acta Autom Sinica 35(6):717–724
9. Kusiak A (2001) Rough set theory a data mining tool for semiconductor manufacturing. IEEE Trans Electron Pack Manuf 24(1):44–50
10. Kusiak A (2000) Decomposition in data mining an industrial case study. IEEE Trans Electron Pack Manuf 23(4):345–353
11. Kusiak A (2001) Feature transformation methods in data mining. IEEE Trans Electron Pack Manuf 24(3):214–221
12. Chen YM, Miao DQ, Wang RZ (2010) A rough set approach to feature selection based on ant colony optimization. Pattern Recognit Lett 31:226–233
13. Shiue YR, Su CT (2002) Attribute selection for neural network based adaptive scheduling systems in flexible manufacturing systems. Int J Adv Manuf Technol 20:532–544
14. Shiue YR, Guh RS (2006) The optimization of attribute selection in decision tree-based production control systems. Int J Adv Manuf Technol 28:737–746
15. Shiue YR, Guh RS (2006) Learning based multi pass adaptive scheduling for a dynamic manufacturing cell environment. Robot Comput-Integr Manuf 33:203–216
16. Shiue YR (2009) Development of two-level decision tree-based real-time scheduling system under product mix variety environment. Robot Comput-Integr Manuf 25:709–720
17. Shiue YR, Guh RS, Tseng TY (2009) A GA based learning bias selection mechanism for real time scheduling systems. Expert Syst Appl 36:11451–11460
18. Hu CH, Su SF (2004) Hierarchical clustering methods for semiconductor manufacturing data. In: Proceedings of the 2004 IEEE international conference on networking, sensing control, Taiwan, 2004, pp 1063–1068
19. Chen T (2007) Predicting wafer-lot output time with a hybrid FCM–FBPN approach. IEEE Trans Syst Man Cybern Part B: Cybern 37(4):784–793
20. Chen T (2007) An intelligent hybrid system for wafer lot output time prediction. Adv Eng Inform 21:55–65
21. Koonce DA, Tsai SC (2000) Using data mining to find patterns in genetic algorithm solutions to a job shop schedule. Comput Ind Eng 38:361–374
22. Li XN (2006) Application of data mining in scheduling of single machine system. Ph.D. dissertation, Iowa State University, USA
23. Rafinejad SN, Ramtin F, Arabani AB (2009) A new approach to generate rules in genetic algorithm solution to a job shop schedule by fuzzy clustering. In: Proceedings of the world congress on engineering and computer science, USA
24. Han J, Kamber M, Pei J (2006) Data mining: concepts and techniques. Morgan Kaufmann
25. Liu Y (2007) Research and application of statistical method of data reduction. Xiamen University
26. Lakshminarayan K, Harp SA, Goldman RP et al (1996) Imputation of missing data using machine learning techniques. KDD, pp 140–145
27. Royston P (2004) Multiple imputation of missing values. Stata J 4:227–241
28. Nelwamondo FV, Mohamed S, Marwala T (2007) Missing data: a comparison of neural network and expectation maximisation techniques. arXiv preprint arXiv:0704.3474
29. Wang X, Deng X, Liu Y et al (2012) A method for missing data interpolation by SVR. In: 2012 IEEE symposium on electrical & electronics engineering (EEESYM). IEEE, pp 132–135
30. García-Laencina PJ, Sancho-Gómez JL, Figueiras-Vidal AR et al (2009) K nearest neighbours with mutual information for simultaneous classification and missing data imputation. Neurocomputing 72(7) 1483–1493
31. Keerin P, Kurutach W, Boongoen T (2012) Cluster-based KNN missing value imputation for DNA microarray data. In: 2012 IEEE international conference on systems, man, and cybernetics (SMC). IEEE, pp 445–450

32. Dawkins R (1976) The selfish gene. Oxford University Press, New York
33. Moscato (1989) On evolution, search, optimization, GAs and martial arts: toward memetic algorithms. California Inst. Technol., Pasadena, CA, Tech. Rep. Caltech Concurrent Comput. Prog. Rep. 826
34. Kennedy J, Eberhart RC (1995) Particle swarm optimization. In: Proceedings of IEEE international conference on neural networks, vol IV, pp 1942–1948
35. Dorigo M, Maniezzo V, Colorni A (1996) Ant system: optimization by a colony of cooperating agents. IEEE Trans Syst Man Cybern Part B 26(1):29–41
36. Wolpert DH, Macready WG (1997) No free lunch theorems for optimization. IEEE Trans Evol Comput 1:67–82
37. Liu B, Wang L, Jin YH (2007) An effective PSO-based memetic algorithm for flow shop scheduling. IEEE Trans Syst Man Cybern Part B: Cybern 37(1):18–27
38. Smith JE (2007) Coevolving memetic algorithms: a review and progress report. IEEE Trans Syst Man Cybern Part B: Cybern 37(1):6–17
39. Ong YS, Lim MH, Chen XS (2010) Research frontier: memetic computation—past, present & future. IEEE Comput Intell Mag 5(2):24–36
40. Iacca G, Neri F, Minino E, Ong Y-S, Lim M-H (2012) Ockham's razor in memetic computing: three stage optimal memetic exploration. Inf Sci 188:17–42
41. Kelly JD (1991) A hybrid genetic algorithm for classification. In: Proceedings of the 12th international joint conference on artificial intelligence, pp 645–650
42. JT Ren, XL Zhuo, XL Xu, SC Yin (2007) PSO based feature weighting algorithm for KNN. Comput Sci 34(5) (in Chinese)
43. Merwe DW, Engelbrecht AP (2003) Data clustering using particle swarm optimization. In: IEEE congress on evolutionary computation, pp 215–220
44. MacQueen J (1967) Some methods for classification and analysis of multivariate observations. In: Proceedings of 5th Berkeley symposium on mathematical statistics and probability. University of California Press, pp 281–297
45. Jolliffe I (2005) Principal component analysis. Wiley
46. Sarle WS (1990) The VARCLUS procedure. SAS/STAT user's guide. SAS Institute Inc., Cary, NC, USA

Chapter 4
Correlation Analysis of Performance Index of Semiconductor Production Line

Semiconductor production line is a typical complex manufacturing system with numerous processing equipment and extremely complex process flow. There are usually more than a dozen types of products being processed at the same time on the same production line, which makes the competition for the right to use online equipment in products more intense. The scheduling scheme and dispatching strategy of semiconductor production line will greatly affect the current production line working conditions, and the quality of scheduling strategy will directly affect the queue length of each equipment and the waiting time of each card workpiece, and then affect the operation efficiency of the whole production line from a global perspective. According to different orders, product demand combinations with different types, quantities and urgency will be produced in the production line, which makes the bottlenecks in the specific production implementation different, and the production process is highly uncertain. The above factors make the information redundancy of online production data high, and the inline relationship is very complex. This also makes the relationship between performance indexes calculated based on daily production data become complicated.

4.1 Long-Term and Short-Term Performance Indicators of Semiconductor Manufacturing System

Performance index is based on actual production data statistics, which reflects the evaluation index of current manufacturing system operation efficiency [1–3]. Optimizing performance index is the goal of studying semiconductor production scheduling, that is, improving efficiency and increasing production capacity, so as to make the whole manufacturing system operate efficiently and well [4–6]. Referring to the related work of Qiao et al. [7] in the design of performance index system

L. Li et al., *Data-Driven Scheduling of Semiconductor Manufacturing Systems*, Advanced and Intelligent Manufacturing in China,
https://doi.org/10.1007/978-981-19-7588-2_4

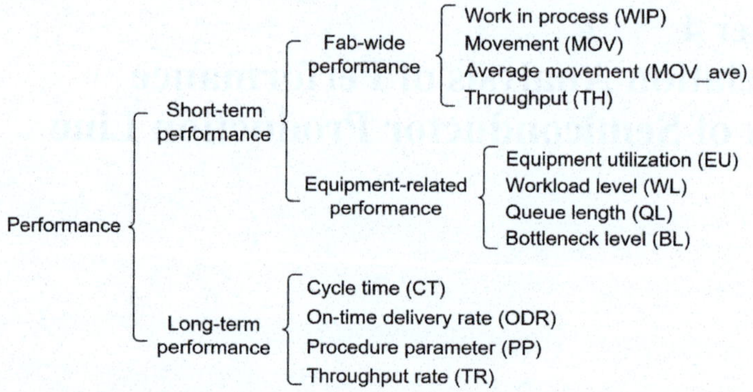

Fig. 4.1 Performance index evaluation system of semiconductor production line

in semiconductor wafer manufacturing system, the performance index evaluation system is classified as follows, as shown in Fig. 4.1.

According to different statistical periods, the performance indexes of semiconductor manufacturing system can be divided into short-term performance indexes and long-term performance indexes [8]. Among them, the short-term performance index is obtained by analyzing and counting the production data in a short period, which can directly and clearly reflect the objective production status of the current production line, reflect the operation efficiency of the production line, and thus reflect the advantages and disadvantages of the daily production planning and scheduling scheme. Long-term performance index refers to the statistics obtained through a long manufacturing cycle, which can comprehensively reflect the implementation effect of the daily releasing plan and scheduling strategy of the current production line [9], which is the most concerned index for enterprises and customers. According to different practical uses, the evaluation indexes can be further subdivided into four categories. Among them, short-term performance indicators mainly include indicators related to production lines, equipment, and workpieces [10]. Long-term performance indicators mainly refer to indicators directly related to products [11].

(1) Performance indicators related to the production line. Whether the scheduling scheme can effectively control the current production line can be quickly reflected by these indicators. It mainly includes daily Work in Process (WIP), daily MOVE steps, sunrise Throughput, daily average move rate (Turn), etc.

    WIP: Refers to the total number of workpieces that have been put into the semiconductor production line and have not completed all the processing steps, that is, the total number of silicon wafers on the production line, with a statistical period of 24 h. Its value is the sum of the number of workpieces waiting to be processed in each processing buffer zone and the number of workpieces being processed on all equipment in each processing zone.

$$W_t = \sum_{j=1}^{n_t} W_{t,j} + W_{t,wait} \tag{4.1}$$

$$W_f = \sum_t W_t \tag{4.2}$$

$W_{t,j}$ refers to the number of work-in-process being processed on equipment j, n_t indicates the total number of equipment in the processing zone t, $W_{t,wait}$ refers to the number of work-in-process waiting to be processed in the buffer zone in the processing zone t, W_t refers to the number of work-in-process in the processing zone t, W_f indicates the total number of work-in-process in the production line.

MOVE every day: Take 24 h as the statistical period to calculate the move steps of all workpieces on the production line. A work piece completes a processing step on a certain equipment, which is called a movement. The higher the daily movement, the more processing tasks completed by the production line. MOVE is an important index to measure the performance of semiconductor production line. The higher its value, the shorter the waiting time of workpiece, the higher the processing capacity of production line and the higher the utilization rate of equipment.

$$\text{Move} = \sum_i \sum_j \sigma_j * P_{i,j} \tag{4.3}$$

Move indicates the daily moving steps of the production line within 24 h. σ_j indicates whether the jth processing of the ith equipment is completed, completed as 1, unfinished as 0. $P_{i,j}$ indicates the number of workpieces processed by the jth processing of the ith equipment.

Sunrise quantity (TH): Refers to the number of workpieces that have completed all processing steps on the production line that day.

$$\text{TH}_{\text{24h}} = \sum_i W_{x=0} \tag{4.4}$$

$W_{x=0}$ indicates the number of workpieces with zero remaining processing steps in the last processing area of the production line, that is, the test area.

Average daily moving speed: Refers to the average number of moving steps of each workpiece within 24 h.

$$v_{\text{24h}} = \text{Move}/W_f \tag{4.5}$$

In which the daily moving steps of the production line are expressed, and the WIP quantity of the current day is expressed.

(2) Equipment-related performance indicators. They reflect the operation of the equipment. Semiconductor manufacturing is a capital-intensive industry, so producers pursue efficient utilization of equipment, including Equipment Utility and equipment queue leader. Among them, the equipment utilization rate reflects the actual operation efficiency of the system and is the most important performance index related to equipment.

Average daily equipment utilization rate (EQU_UTI): Refers to the ratio of the time used for processing workpieces of a certain equipment to the total startup time of the day, taking 24 h as the statistical period.

$$P_u = \sum_{i=1}^{m} \frac{T_{ih}}{T_{op}} * 100\% \quad (4.6)$$

Among them, P_u refers to the utilization rate of equipment u, T_{ih} the use time of the equipment's first operation i, m the total operation times of the equipment on the same day, and T_{op} the startup time of the same day.

Daily queue leader (QL): Refers to the 24-h statistical period which is used to calculate the number of all the workpieces waiting to be processed in the corresponding processing buffer in the current production line, $W_{t,wait}$ as shown in Formula (4.1).

(3) Workpiece-related performance indicators. They reflect the process flow and processing situation of each wafer in the whole life cycle of the production line, mainly including the Waiting Time of each wafer, the current remaining processing steps, the delivery date, whether it is an urgent workpiece and other information.

Workpiece waiting time sum (WT): Refers to the sum of waiting time in all processing buffers after a workpiece is put into the production line.

$$WT = \sum_{i=1}^{n} t_i \quad (4.7)$$

n indicates the total number of processing zones, and t_i indicates the time waiting for processing in the buffer of the ith processing zone.

The total waiting time of the workpiece in the buffer zone is an objective embodiment of the variable cost in semiconductor manufacturing, which intuitively reflects the wasted time of the workpiece in the whole production line. In actual production, it can be observed that the completion events of silicon wafers are discrete and non-uniform, because the waiting time of silicon wafers in the whole life cycle of processing is discrete. The length of processing time depends on the congestion degree of each bottleneck equipment area of the current production line, that is, the queuing time of each card workpiece. The generation and congestion degree of bottleneck equipment area depend on the current scheduling strategy, the frequency of access to bottleneck equipment by all kinds of products and the usage time required by the process. The congestion degree determines the waiting time of the workpiece in a certain equipment area. Therefore, waiting time is the comprehensive result of the current production scheme, and it is an important performance index related to workpieces, which is directly or indirectly affected by other short-term performance indexes.

(4) Product-related performance indicators. This is the performance index of semiconductor production line directly related to the final product, mainly including Cycle Time and on-time delivery rate, etc.

Processing cycle (CT): Refers to the time required for silicon wafers from being put into production line to completing all processing steps.

$$CT_i = t_{i,out} - t_{i,in} \quad (4.8)$$

Among them, CT_i indicates the processing cycle of the workpiece i, $t_{i,out}$ the time when the workpiece i completes all processing procedures, and $t_{i,in}$ the time when the workpiece i enters the production line and starts to prepare for processing.

The average processing cycle of wafers is long, usually ranging from 1 to 3 months. Accurately grasping the processing cycle is the key for enterprises to maintain competitiveness. Semiconductor manufacturing system has the characteristics of large scale, complex processing technology and multi-entry process, which makes its processing cycle not only depend on its own process requirements, but also depends on the current scheduling scheme. That is, the processing cycle will change with the change of real-time working conditions of the current production line.

Just-in-time delivery rate (ODR): It reflects the completion degree of the wafer processing factory to the production task. It usually takes longer manufacturing cycle to get statistics, which is a long-term performance of the advantages and disadvantages of scheduling schemes.

$$ODR_{z,T_d = \frac{n_1}{n_1+n_2}} \quad (4.9)$$

Among them, ODR_{z,T_d} refers to the on-time delivery rate of z class products in the cycle T_d, which n_1 indicates the number of all on-time delivered workpieces in class z products and n_2 the number of all undelivered workpieces in class z products.

Performance index is the evaluation index of scheduling scheme and dispatching rule in semiconductor manufacturing system. Usually, the fluctuation of these indicators can quickly reflect the change of scheduling rules. From the factory point of view, they are valuable data which are easy to collect and can directly reflect the status of the production line. Because the data of the production line are numerous and fine, the higher the digitization degree, the higher the collection frequency and finer granularity of data points in the factory, which not only brings more complete and detailed data, but also makes the inline relationship of short-term performance indicators more complex and the coupling degree between data is higher, which makes it difficult to quantify the mathematical relationship among performance indicators.

In addition, there are inevitably some constraints between the long-term and short-term performance indicators, so the above indicators reflecting the operational performance of semiconductor production lines cannot be optimized at the same time [12]. All kinds of scheduling schemes are designed and optimized to achieve the tradeoff and balance between performance indexes [13]. For example, if the average processing cycle of wafers is to be shortened, the work-in-process level of the production line should be reduced, so that the waiting time of workpieces can be reduced. Reducing the number of work-in-process can reduce the production and operation costs of factories, and at the same time can effectively improve the yield. However, if the WIP level is too low, the equipment utilization rate of the production line will be significantly reduced, thus affecting the daily moving steps and productivity [14]. The reduction of production efficiency will greatly weaken the profitability of enterprises and lead to the growth of capital withdrawal cycle. On the contrary, if the work-in-process level is too high, although the equipment utilization rate and daily moving steps are improved, it may reduce the moving speed of the production line, increase the average processing cycle instead, reduce the yield, and reduce the liquidity of enterprise funds, affecting the profitability of the factory [15]. The balance among performance indexes is what a good scheduling scheme should pursue. On this basis, we should pay attention to the optimization of some key performance indexes to make the overall performance of the production line reach the global approximate optimum [16]. Therefore, the mathematical modeling of the intrinsic correlation of performance indicators can quantify the constraint relationship between indicators, so as to pay more attention to some key indicators in the design of scheduling scheme and obtain the global optimal effect [17].

4.2 Statistical Analysis of Performance Indicators for Semiconductor Production Lines

In this paper, the historical data of a semiconductor manufacturing enterprise will be taken as the object, and the statistics and correlation analysis of performance indicators will be carried out. The data sample is the production line data from January to December 2013. 31 types of production data are collected every 4 h, and exported as csv files, which contain 6 data sets every day. The production data set covers almost all the information of the current production line, among which the actual conditions of the production line mainly include WIP information table, equipment information table, moving history information table and data collection schedule. The statistics of performance indicators concerned in this paper are mainly based on these four tables. Table 4.1 lists the key parameter information used in each table. Among them, the WIP information table provides the basic information of WIP in the current production line, which is presented in the form of flow information. The equipment information table covers equipment process information related to processing. The movement history information table records the movement history of the workpiece on the production line. The data collection schedule records the collection time of this data set.

In this paper, for a certain type of production data, such as work-in-process information table, all the collected production data information is merged into a table in time sequence, covering the annual information of the production line indicators. Then, according to the statistical method of performance index, the required long-term and short-term performance index data are obtained.

Table 4.1 Production data information table

Data table name	Parameter
Work in process information table t_wip.csv	Workpiece card number of this card
	Version number of this card workpiece
	Number of wafers contained in the card workpiece
	The current station number of the card workpiece
	The large group number of the process being executed by the workpiece of this card
	The card type of the card workpiece
	Card status
	Number of remaining steps
	Contract delivery date of this card workpiece
	Time when the card workpiece enters the production line
Equipment information table t_eqipment.csv	Description information of the device
	Classification of processing zone where equipment is located
	Equipment processing capacity
The move history information table t_move_history.csv	Card number
	Equipment number
	Site
	Processing menu
	Workpiece entry and exit time
	Workpiece move-in date
Current data collection schedule t_time.csv	Data acquisition time

4.2.1 Short-Term Performance Indicators

(1) WIP (piece/day)

Combine the six time data tables sampled every day according to the table name t_wip.csv. Screen under the attribute "lot state" of each table, and select the cards whose "card state" is "processing" and "waiting", and remove the duplicate card numbers. Count the number of WIP pieces per day.

(2) The number of daily moving steps (MOVE, steps/day)

Combine the six-time data packets sampled every day according to the table name t_wip.csv, filter under the attribute "lot state" of each table and select the card flow information whose "card state" is "processing". Each MOVE means that a workpiece is processed every time it completes a working procedure, and it is counted by day.

Figure 4.2 is the annual trend diagram of the relationship between WIP as independent variable and MOVE as dependent variable, in which the dashed line fits the linear trend between them. The left diagram is fitted according to time, and it shows the relationship between the annual WIP, and the daily moving steps MOVE in time sequence. The right figure does not consider the time sequence, but only studies the change trend of daily moving steps with the increase of WIP. According to the trend line, MOVE basically increases monotonously with WIP. At the beginning of the year, the production line was restarted, and no new pieces were put into operation, so the number of WIP was very low.

(3) Queue Length (pieces/day)

T_wip.csv is extracted from six time data sampled every day, and the card whose card status is "Waiting" is selected by filtering under the attribute "Lot State" of each table. For the queue leader, this paper counts the queue leader once for every six T_WIP.csvs every day, and then takes the average as the queue leader of the day. The annual change of MOVE with QL quantity is shown in Fig. 4.3.

Figure 4.3 is an annual trend diagram of the relationship between the daily queue length QL as an independent variable and the daily moving steps MOVE as

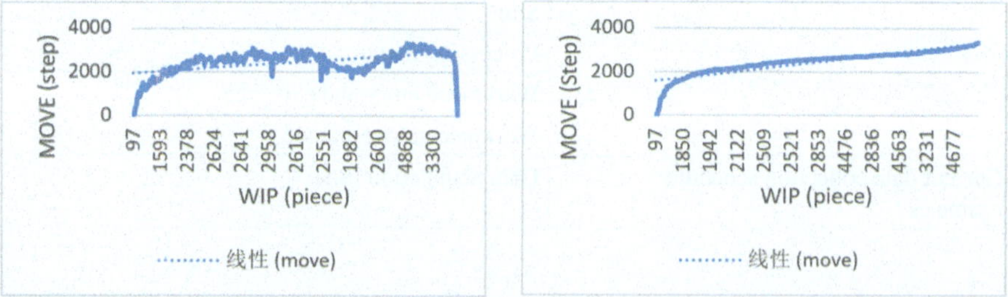

Fig. 4.2 Trend diagram of Move_WIP relationship

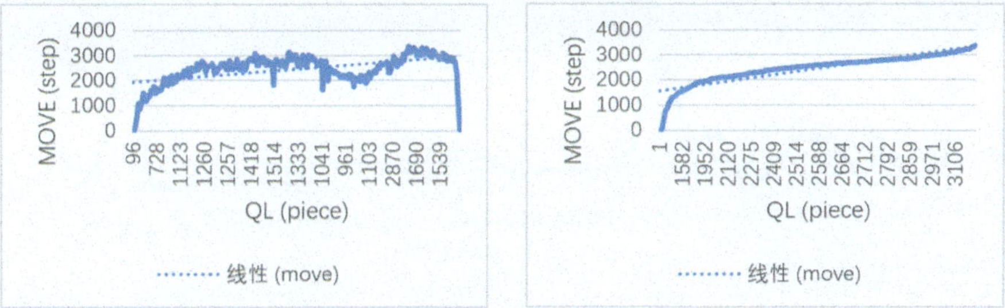

Fig. 4.3 Trend diagram of Move_QL relationship

a dependent variable, in which the dashed line fits the linear trend of the change between them. Similar to Fig. 4.2, the left diagram is listed in chronological order, while the right diagram does not consider chronological order, and only studies the changing trend of daily moving steps with the increase of the number of WIP. It can be seen that with the increase of daily queue length, the trend of daily moving steps also becomes larger. Only when the queue length increases, the increasing trend slows down, which shows that the production line is not overloaded because of the increase of queue length, but only reduces the moving speed.

Figure 4.4 shows the daily WIP quantity and daily queue length of the whole year, in which the blue line represents the trend of WIP quantity, and the orange line represents the trend of daily queue length. It can be found from the figure that the trends of the two are almost the same, and the difference between the two is basically stable. The difference between the two indicates the number of pieces being processed on the equipment every day, which shows that the number of pieces being processed every day in most of the year is relatively stable.

(4) Equipment utilization rate (EQI_UTI)

About equipment utilization rate, the time when a workpiece enters and exits a certain equipment can be found in t_move_history.csv, and the equipment utilization rate of the current day can be calculated for each equipment, that is,

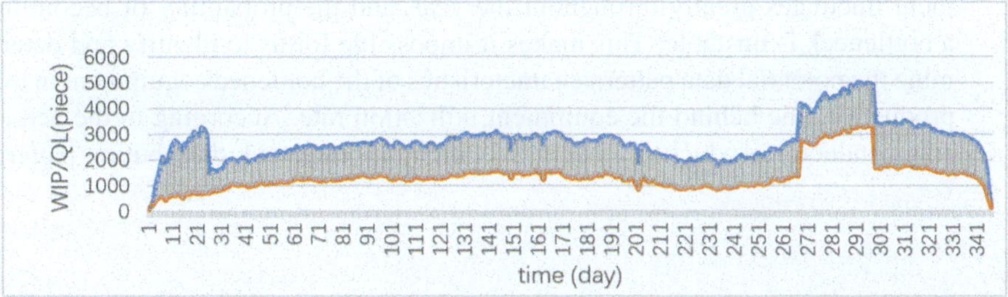

Fig. 4.4 Annual WIP quantity and daily queue length

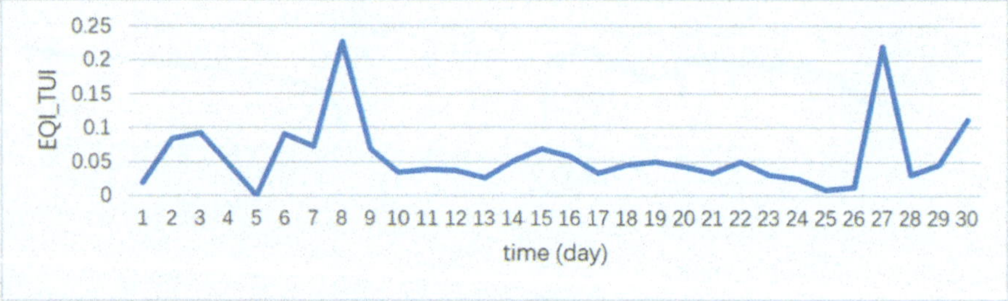

Fig. 4.5 Utilization trend of a certain equipment in June

the equipment number is used as identification for statistics. Figure 4.5 is the trend chart of equipment utilization rate of a certain equipment in June. It can be seen from the figure that the utilization rate of equipment is unstable, and changes greatly with time in one month.

In many manufacturing systems, the potential data model of bottleneck equipment distribution in the production line is single, that is, all bottleneck links in the production line can be abstractly simplified into a production model with a single bottleneck node. In other manufacturing systems, because of the scattered existence of bottleneck equipment and the unstable occurrence time of bottleneck equipment, the system can not be simplified into a single bottleneck production model. This also means that the internal relations of production data under different production models are different, so modeling methods should be designed according to the characteristics of production models.

Define the bottleneck equipment when the monthly utilization rate of equipment is more than 60%. Figure 4.6 selects three equipments with high correlation with daily moving steps, showing that they become bottlenecks of production lines all year round. 1 means that the monthly average utilization rate of the equipment is more than 60% in this month, which becomes the bottleneck equipment. 0 means that the monthly average utilization rate of the equipment is less than 60% in this month, and it is a non-bottleneck equipment. Through the direct observation of the equipment utilization rate, it is found that the equipment utilization rate on the production line is unstable throughout the year, that is, it is not always the bottleneck equipment. The utilization rate of equipment fluctuates greatly throughout the year, and the probability of becoming a bottleneck is unstable. This makes it impossible for us to identify and determine the potential data pattern characteristics of the bottleneck equipment in the production line behind the equipment utilization rate. According to the actual semiconductor production system, its potential bottleneck distribution pattern must conform to one of the above two production models. Therefore, when forecasting and modeling performance indicators, the forecasting methods should be distinguished between single bottleneck and multi-bottleneck production models.

(5) Waiting time of workpiece for processing

The waiting time of a workpiece is obtained by the difference between the residence time of a workpiece in the production line (i.e. the processing cycle) and the processing time on the equipment. Through the information statistics of each card workpiece in the whole production life cycle, we can know the access time of the workpiece in all the equipment used, and we can get the actual processing time of the workpiece by adding them up. Figure 4.7 is a statistical diagram of the average waiting time of each product version on the production line.

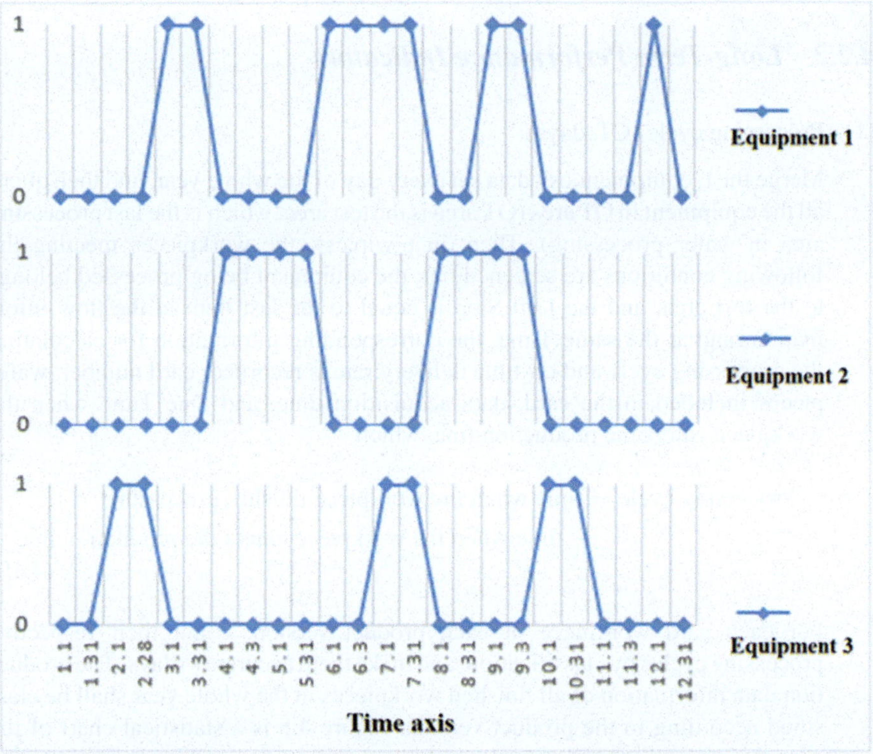

Fig. 4.6 The situation of local equipment becoming the bottleneck of production line in the whole year

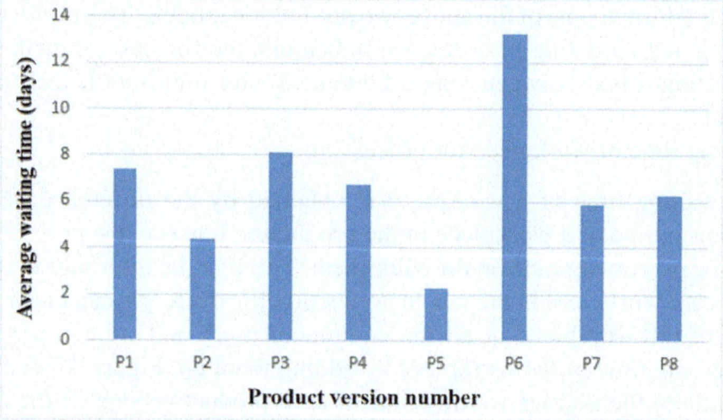

Fig. 4.7 Statistical diagram of average waiting time of each product version on the production line

4.2.2 Long-Term Performance Indicators

(1) Processing cycle (CT, days)

Merge the t_equipment.csv data set every day of the whole year, and then query all the equipment in OT area (OT area is the test area, which is the last processing area in wafer processing). Then, in t_wip.csv, the workpieces meeting the following conditions are screened out: the equipment being processed belongs to the test area, and the Left Step is equal to the last item in the flow information, and at the same Time, the corresponding information for calculating the processing cycle and on-time delivery rate is recorded: card number, wafer pieces included in the card, data acquisition time, and Due Time when the workpiece enters the production line, which

$$\text{Processing cycle} = \text{time when the workpiece finishes processing} \\ - \text{time when the workpiece enters the production line} \tag{4.10}$$

For each card workpiece of each product version, count their respective processing cycles by classification, and make them accurate to days. The production data information of all finished workpieces in the whole year shall be classified according to the product version. Figure 4.8 is a statistical chart of the average processing time of each product version.

(2) Throughput (throughput)

According to the above, the completion card information has been obtained, such as the card completion time (i.e., the data acquisition time of the work-in-process flow information of the card), and the number of wafers contained in

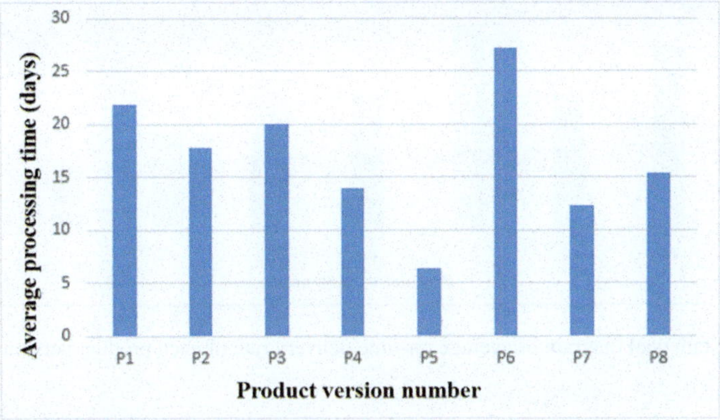

Fig. 4.8 Statistical diagram of average processing time of each product version

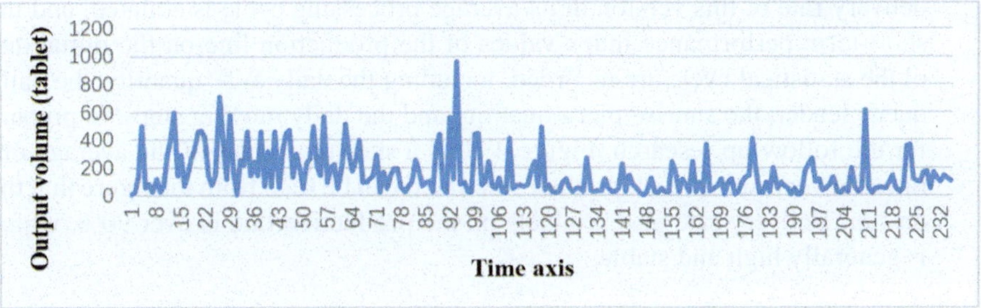

Fig. 4.9 Statistical chart of annual film output of the production line

the card, which is counted by day. Figure 4.9 is a statistical chart of the annual output of the production line.

(3) On-time delivery rate

The judgment is also made according to the completion card information obtained above. If the data acquisition time is less than or equal to the contract due time of the card workpiece, it is judged that the card workpiece is delivered on time; If the data acquisition time (time) is greater than the contract delivery time (due time) of the card workpiece, it is determined that the card workpiece is delayed.

$$\text{On-time delivery rate} \\ = \text{number of on-time delivery cards/number of all completed cards} \tag{4.11}$$

For each version of the product, if its average processing cycle is taken as the statistical cycle of the on-time delivery rate of this version, there will be

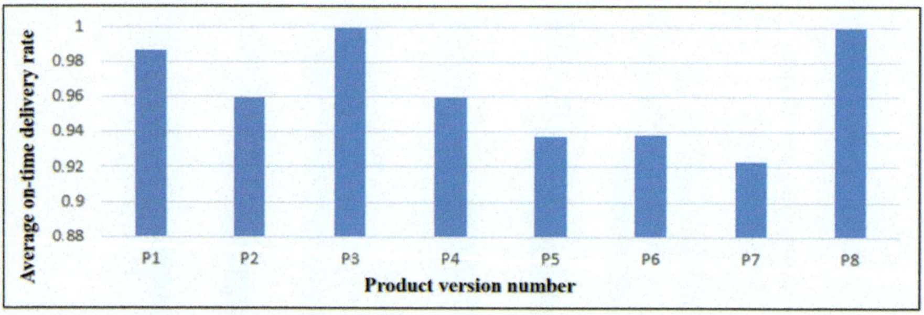

Fig. 4.10 Statistical diagram of average on-time delivery rate of each product version

less statistical data of the on-time delivery rate, so the training set is generated by rolling. The rolling cycle is 1 day, and the rolling window is the average processing cycle of this version of products. Every other day, the on-time delivery rate of this version in its average processing cycle is counted, and the short-term performance index values of the production line on the initial day of the statistical cycle are recorded, including the daily WIP quantity, the daily queue leader, the sunrise piece quantity and the daily moving steps, to prepare for the follow-up research. Figure 4.10 is a statistical chart of the average on-time delivery rate of each product version. It can be seen from the figure that the on-time delivery rate of 8 products on the production line is over 90%, which is generally high and stable.

4.3 Correlation Analysis of Performance Indicators Based on Correlation Coefficient Methods

There are many performance indicators in a production line, so it is necessary to study the correlation between them. This section uses the correlation coefficient analysis method to analyze the correlation between them.

The principle of correlation coefficient analysis method is described as follows:

For matrices A and B, the correlation coefficient between them is:

$$R = \frac{C(B, A)}{\sqrt{C(B, B)C(A, A)}}$$

where $C = \text{cov}(A, B)$ refers to the covariance of matrix A and B.

The value of R is between [− 1, 1], where “1” represents the maximum positive correlation between matrix A and B, and “− 1” represents the maximum negative correlation between them.

This chapter will simulate two actual production lines to obtain corresponding performance index data, and then analyze the correlation of performance index.

(1) A semiconductor production line (BL6): the production line contains 119 equipments, which are divided into 19 work areas (according to equipment functions), and there are 10 kinds of workpieces on the production line, which have five dispatching rules of FIFO, EDD, SPT, LS and CR and three working conditions of WIP of 6000 (light load), 7000 (full load) and 8000 (overload), respectively.
(2) Standard semiconductor production line MIMAC: The production line contains 229 equipments, which are divided into 104 equipment groups (according to replaceable equipment, all the equipments in one equipment group are replaceable equipment), and there are 9 kinds of workpieces on the production line. Under five dispatching rules of FIFO, EDD, SPT, LS and CR, and five working conditions of WIP of 4000 pieces, 5000 pieces, 6000 pieces, 7000 pieces and 8000 pieces respectively, a total of 25 cases were carried out, which was equivalent to the actual situation for 90 days, and many data were obtained. The data of warm-up period in the first 30 days of the production line were removed, and the ones after the production line was stable (the last 60 days) were taken.

All short-term performance indicators in the production line are obtained daily, that is, they are counted once a day. Long-term performance indicators are counted once every ten days.

4.3.1 Block Diagram of Correlation Analysis

As shown in Fig. 4.11, we designed a block diagram for processing simulation data by correlation coefficient analysis. The data obtained from MIMAC of semiconductor production line (BL6) and standard semiconductor production line are processed by this block diagram flow.

The main flow of block diagram is as follows:

① The treatment process is divided into three parts, the first two parts are for different working conditions and dispatching rules, focusing on short-term performance indicators, and the third part is to consider the relationship between long-term performance indicators and short-term performance indicators;
② The correlation coefficient method is used to obtain the correlation of each block;
③ Carry out the correlation analysis of the whole production line including the combination of working conditions and dispatching rules;
④ Get the correlation between long-term and short-term performance indexes;
⑤ Establish performance index system.

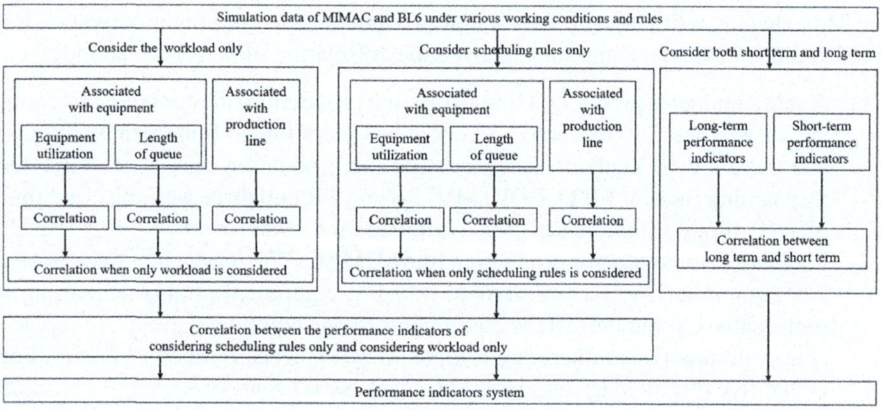

Fig. 4.11 Block diagram of data processing by correlation coefficient method

4.3.2 Correlation Analysis of Performance Indicators Considering Working Conditions

In this study, we selected equipment with daily average equipment utilization rate greater than 55% for correlation analysis, and the selected equipment is shown in Table 4.2.

As shown in Table 4.2, the selected 11 equipments are distributed in three areas, which are lithography area, dry etching area and wet etching area.

(1) Correlation analysis of equipment related performance indicators

According to the 11 selected equipments, the equipment utilization rate under different working conditions (light load, full-load and overload) is counted, and the correlation coefficient between them is calculated by using the correlation

Table 4.2 Selected devices and their corresponding workspaces

Workspace	Device name
Lithography area	BL_6TELC1
	BL_6STP08
	BL_6TELC2
	BL_6TELD1
	BL_6STP09
	BL_6ADI01
	BL_6TELD2
Dry engraving area	BL_6GAN03
	BL_6OVN10
Wet engraving area	BL_6WET22
	BL_6WET21

coefficient method. The results are expressed in matrix form, as shown in (4.12), (4.13) and (4.14) respectively.

$$A = \begin{bmatrix} a_{11} & \cdots & a_{1n} \\ \vdots & \ddots & \vdots \\ a_{111} & \cdots & a_{1111} \end{bmatrix} = \begin{bmatrix} 1 & 0 & 0 & 0 & 0 & 0 & 0 & 0 & 0 & 0 & 0 \\ 0.71 & 1 & 0 & 0 & 0 & 0 & 0 & 0 & 0 & 0 & 0 \\ 0.92 & 0.69 & 1 & 0 & 0 & 0 & 0 & 0 & 0 & 0 & 0 \\ 0.81 & 0.77 & 0.80 & 1 & 0 & 0 & 0 & 0 & 0 & 0 & 0 \\ 0.83 & 0.74 & 0.84 & 0.86 & 1 & 0 & 0 & 0 & 0 & 0 & 0 \\ 0.77 & 0.66 & 0.77 & 0.88 & 0.79 & 1 & 0 & 0 & 0 & 0 & 0 \\ 0.83 & 0.73 & 0.85 & 0.90 & 0.87 & 0.90 & 1 & 0 & 0 & 0 & 0 \\ 0 & 0 & 0 & 0 & 0 & 0 & 0 & 1 & 0 & 0 & 0 \\ 0 & 0 & 0 & 0 & 0 & 0 & 0 & 0.81 & 1 & 0 & 0 \\ 0 & 0 & 0 & 0 & 0 & 0 & 0 & 0 & 0 & 1 & 0 \\ 0 & 0 & 0 & 0 & 0 & 0 & 0 & 0.55 & 0 & 0.75 & 1 \end{bmatrix} \tag{4.12}$$

where a_{mn} represents the correlation coefficient between the mth and nth elements. To make the obtained matrix more intuitive, the value less than 0.5 is assigned as 0. The elements of the rows and columns of the matrix represent the devices BL_6TELC1, BL_6STP08, BL_6TELC2, BL_6TELD1, BL_6STP09, BL_6ADI01, BL_6TELD2, BL_6GAN03, BL_6OVN10, BL_6WET22 and bl _ 6wet from top to bottom, respectively e.g., a_{11} equal to 1, which means that the correlation coefficient between the utilization rate of the equipment BL_6TELC1 and its own utilization rate is 1. Another example is a_{21} equal to 0.71, which means that the correlation coefficient of utilization ratio of equipment BL_6STP08 and BL_6TELC1 is 0.71. Unless otherwise specified, the same rules apply.

$$A = \begin{bmatrix} a_{11} & \cdots & a_{1n} \\ \vdots & \ddots & \vdots \\ a_{111} & \cdots & a_{1111} \end{bmatrix} = \begin{bmatrix} 1 & 0 & 0 & 0 & 0 & 0 & 0 & 0 & 0 & 0 & 0 \\ 0.71 & 1 & 0 & 0 & 0 & 0 & 0 & 0 & 0 & 0 & 0 \\ 0.92 & 0.69 & 1 & 0 & 0 & 0 & 0 & 0 & 0 & 0 & 0 \\ 0.81 & 0.81 & 0.80 & 1 & 0 & 0 & 0 & 0 & 0 & 0 & 0 \\ 0.81 & 0.81 & 0.84 & 0.84 & 1 & 0 & 0 & 0 & 0 & 0 & 0 \\ 0.76 & 0.70 & 0.78 & 0.88 & 0.79 & 1 & 0 & 0 & 0 & 0 & 0 \\ 0.80 & 0.77 & 0.82 & 0.92 & 0.86 & 0.91 & 1 & 0 & 0 & 0 & 0 \\ 0 & 0 & 0 & 0 & 0 & 0 & 0 & 1 & 0 & 0 & 0 \\ 0 & 0 & 0 & 0 & 0 & 0 & 0 & 0.80 & 1 & 0 & 0 \\ 0 & 0 & 0 & 0 & 0 & 0 & 0 & 0 & 0 & 1 & 0 \\ 0 & 0 & 0 & 0 & 0 & 0 & 0 & 0.54 & 0 & 0.74 & 1 \end{bmatrix} \tag{4.13}$$

$$A=\begin{bmatrix} a_{11} & \cdots & a_{1n} \\ \vdots & \ddots & \vdots \\ a_{111} & \cdots & a_{1111} \end{bmatrix}=\begin{bmatrix} 1 & 0 & 0 & 0 & 0 & 0 & 0 & 0 & 0 & 0 & 0 \\ 0.71 & 1 & 0 & 0 & 0 & 0 & 0 & 0 & 0 & 0 & 0 \\ 0.92 & 0.69 & 1 & 0 & 0 & 0 & 0 & 0 & 0 & 0 & 0 \\ 0.81 & 0.81 & 0.80 & 1 & 0 & 0 & 0 & 0 & 0 & 0 & 0 \\ 0.81 & 0.81 & 0.84 & 0.84 & 1 & 0 & 0 & 0 & 0 & 0 & 0 \\ 0.76 & 0.70 & 0.78 & 0.88 & 0.79 & 1 & 0 & 0 & 0 & 0 & 0 \\ 0.80 & 0.77 & 0.82 & 0.92 & 0.86 & 0.91 & 1 & 0 & 0 & 0 & 0 \\ 0 & 0 & 0 & 0 & 0 & 0 & 0 & 1 & 0 & 0 & 0 \\ 0 & 0 & 0 & 0 & 0 & 0 & 0 & 0.80 & 1 & 0 & 0 \\ 0 & 0 & 0 & 0 & 0 & 0 & 0 & 0 & 0 & 1 & 0 \\ 0 & 0 & 0 & 0 & 0 & 0 & 0 & 0.54 & 0 & 0.74 & 1 \end{bmatrix} \tag{4.14}$$

Especially pointed out the following two points:

- As a matter of fact, the matrix formed by the correlation coefficient between pairs obtained from the performance index is a symmetric matrix. Here, to be more intuitive, the upper half of the symmetric matrix is assigned as 0.
- The matrix in the performance analysis of equipment and production line is expressed in the form of matrix groups, and each group has three matrices, which respectively represent the correlation coefficients under the working conditions of 6000 pieces, 7000 pieces and 8000 pieces of WIP.

As shown in Matrix (4.14)–(4.14), the correlation coefficient between the utilization rate of equipment located in the same processing area is higher. Taking the equipment in the lithography area as an example, the correlation coefficient of the utilization ratio of seven equipments in the lithography area is relatively large (greater than 0.7) under three working conditions. The same is true of the equipment utilization correlation in the other two areas (wet etching and dry etching).

According to (4.14)–(4.14):

Conclusion 1: There is a great correlation between the utilization rate of equipment in the same processing area. Therefore, it can be defined that the utilization rate of the workspace is the average utilization rate of the equipment in the zone.

(2) Correlation analysis of related performance indexes of production line

In this section, the correlation analysis of WIP, MOV and TP (short-term performance indicators here) is carried out, and the results are shown in matrices (4.15)–(4.17):

$$A=\begin{bmatrix} a_{11} & a_{12} & a_{13} \\ a_{21} & a_{22} & a_{23} \\ a_{31} & a_{32} & a_{33} \end{bmatrix}=\begin{bmatrix} 1 & 0.11 & -0.12 \\ 0.11 & 1 & -0.19 \\ -0.12 & -0.19 & 1 \end{bmatrix} \tag{4.15}$$

$$A = \begin{bmatrix} a_{11} & a_{12} & a_{13} \\ a_{21} & a_{22} & a_{23} \\ a_{31} & a_{32} & a_{33} \end{bmatrix} = \begin{bmatrix} 1 & -0.03 & -0.02 \\ -0.03 & 1 & 0.05 \\ -0.02 & 0.05 & 1 \end{bmatrix} \tag{4.16}$$

$$A = \begin{bmatrix} a_{11} & a_{12} & a_{13} \\ a_{21} & a_{22} & a_{23} \\ a_{31} & a_{32} & a_{33} \end{bmatrix} = \begin{bmatrix} 1 & -0.01 & 0.001 \\ -0.01 & 1 & -0.16 \\ 0.001 & -0.16 & 1 \end{bmatrix} \tag{4.17}$$

Rows and columns of matrices (4.15), (4.16 and (4.17) respectively represent WIP, MOV and TP.

According to the above data, draw the conclusion 2:

Conclusion 2: The correlation coefficient among the three performance indexes under three working conditions is very small, close to 0. It shows that the correlation between them is very low, and they can't replace each other, that is, all three indicators should be used as performance indicators to measure the advantages and disadvantages of scheduling algorithms.

(3) Pairwise correlation between all short-term performance indexes under three working conditions

WIP, MOV, TP, and the pairwise correlation coefficients among lithography area utilization, wet etching area utilization and dry etching area utilization are shown in matrices (4.18)–(4.20).

$$A = \begin{bmatrix} a_{11} & \cdots & a_{16} \\ \vdots & \ddots & \vdots \\ a_{61} & \cdots & a_{66} \end{bmatrix} = \begin{bmatrix} 1 & 0 & 0 & 0 & 0 & 0 \\ 0 & 1 & 0 & 0 & 0 & 0 \\ 0 & 0 & 1 & 0 & 0 & 0 \\ 0 & 0.69 & 0 & 1 & 0 & 0 \\ 0 & 0 & 0 & 0 & 1 & 0 \\ 0 & 0 & 0 & 0 & 0 & 1 \end{bmatrix} \tag{4.18}$$

$$A = \begin{bmatrix} a_{11} & \cdots & a_{16} \\ \vdots & \ddots & \vdots \\ a_{61} & \cdots & a_{66} \end{bmatrix} = \begin{bmatrix} 1 & 0 & 0 & 0 & 0 & 0 \\ 0 & 1 & 0 & 0 & 0 & 0 \\ 0 & 0 & 1 & 0 & 0 & 0 \\ 0 & 0.67 & 0 & 1 & 0 & 0 \\ 0 & 0 & 0 & 0 & 1 & 0 \\ 0 & 0.53 & 0 & 0 & 0 & 1 \end{bmatrix} \tag{4.19}$$

$$A = \begin{bmatrix} a_{11} & \cdots & a_{16} \\ \vdots & \ddots & \vdots \\ a_{61} & \cdots & a_{66} \end{bmatrix} = \begin{bmatrix} 1 & 0 & 0 & 0 & 0 & 0 \\ 0 & 1 & 0 & 0 & 0 & 0 \\ 0 & 0 & 1 & 0 & 0 & 0 \\ 0 & 0.68 & 0 & 1 & 0 & 0 \\ 0 & 0 & 0 & 0 & 1 & 0 \\ 0 & 0 & 0 & 0 & 0 & 1 \end{bmatrix} \tag{4.20}$$

As shown by matrices (4.18)–(4.20), the correlation coefficient between MOV and lithography area is very high, close to 0.7. This leads to the conclusion 3: The values are all very high, close to 0.7.

Conclusion 3: The positive correlation between MOV and the utilization ratio of lithography area is very large, so the influence of the utilization ratio of lithography area on MOV must be considered when optimizing MOV in future scheduling.

4.3.3 Correlation Analysis of Performance Indicators Considering Dispatching Rules

The processing steps in this subsection are the same as those in Sect. 4.3.2. The processing results of equipment-related performance index correlation and production line-related performance index correlation in this section are consistent with parts (1) and (2) in Sect. 4.3.2. The result matrix is not given here. Come to conclusion 4.

Conclusion 4: Under different dispatching rules, the equipment utilization rate in the same processing zone has a great correlation. Therefore, it can be defined that the utilization rate of the workspace is the average of the equipment utilization rates in the zone.

The correlation coefficient among the three performance indicators under different dispatching rules is very small, close to 0. It shows that the correlation between them is very low, and they can't replace each other. Therefore, these three indicators should be used as performance indicators to measure the advantages and disadvantages of scheduling algorithms.

Then consider the pairwise correlation results of all short-term performance indicators (WIP, MOV, TP, lithography area utilization rate, wet etching area utilization rate and dry etching area utilization rate) under different dispatching rules as shown in Matrix (4.21).

$$A = \begin{bmatrix} a_{11} & \cdots & a_{16} \\ \vdots & \ddots & \vdots \\ a_{61} & \cdots & a_{66} \end{bmatrix} = \begin{bmatrix} 1\,0\,0\,0\,0\,0 \\ 0\,1\,0\,0\,0\,0 \\ 0\,0\,1\,0\,0\,0 \\ 0\,0\,0\,1\,0\,0 \\ 0\,0\,0\,0\,1\,0 \\ 0\,0\,0\,0\,0\,1 \end{bmatrix} \tag{4.21}$$

There should be five matrices here, which represent the results under different dispatching rules of FIFO, EDD, SPT, LS and CR. However, the result matrix under each dispatching rule is consistent, so only one matrix is listed here.

From the matrix (4.21), we can draw the conclusion 5:

Conclusion 5: The correlation among the six performance indexes is less than 0.5. It means that the pairwise correlation between them is very weak and can be ignored. Therefore, there is no situation in which they can be replaced in pairs or have a great influence between pairs. This also means that in the future scheduling optimization (only considering the dispatching rules), if the utilization rate of the area is taken as the measurement standard, the utilization rates of the lithography area, the wet etching area and the dry etching area should be considered separately, and the WIP, MOV and TP values have no direct influence on these three indexes.

4.3.4 Correlation Analysis of Performance Indicators Considering Working Conditions and Dispatching Rules

In this section, three working conditions and five dispatching rules are combined to analyze the correlation of all performance indexes of the whole production line, and the results are shown in Matrix (4.22):

$$A = \begin{bmatrix} a_{11} & \cdots & a_{16} \\ \vdots & \ddots & \vdots \\ a_{61} & \cdots & a_{66} \end{bmatrix} = \begin{bmatrix} 1\,0\,0\,0\,0\,0 \\ 0\,1\,0\,0\,0\,0 \\ 0\,0\,1\,0\,0\,0 \\ 0\,0\,0\,1\,0\,0 \\ 0\,0\,0\,0\,1\,0 \\ 0\,0\,0\,0\,0\,1 \end{bmatrix} \tag{4.22}$$

From matrix (4.22), conclusion 6 can be drawn:

Conclusion 6: The correlation among the six performance indexes is less than 0.5. It means that the pairwise correlation between them is very weak and can be ignored. This also means that there are no cases where they can be replaced by pairs or have great influence between pairs. Therefore, in future scheduling optimization, the consideration is similar to conclusion 5.

4.3.5 Correlation Analysis Between Long-Term Performances and Short-Term Performances

This section analyzes the correlation between short-term performance index MOV and long-term performance index CT, TP and ODR.

Only the correlation results between MOV and CT, TP and ODR under three working conditions are considered: with the increase of working conditions, the correlation coefficients of MOV and CT are 0.52, 0.50 and 0.45 respectively. Other correlation coefficients are close to zero.

Only considering the correlation results between MOV and CT, TP and ODR under five dispatching rules, the correlation between MOV and long-term indicators is different under different dispatching rules. Under FIFO and LS rules, MOV and CT have strong correlation. Under LS, MOV has a strong correlation with TP.

Considering three working conditions and five dispatching rules, there is a negative correlation between CT and ODR.

After analyzing the correlation between long-term performance index and short-term performance index, the conclusion 7:

Conclusion 7: The relationship between MOV and CT is affected by the working conditions. CT was negatively correlated with ODR.

4.3.6 Correlation Analysis of Performances on Production Line Named MIMAC

(1) Correlation analysis of equipment utilization rate

A total of 19 devices with device utilization rate higher than 0.8 under MIMAC model were selected (Table 4.3).

The related coefficient matrix of equipment utilization rate is shown in (4.23):

Table 4.3 Equipment with MIMAC equipment utilization rate greater than 0.8

Equipment	Equipment
WC10123_DNS_3_1	WC13621_IPC_3200_1
WC10123_DNS_3_2	WC13621_IPC_3200_2
WC11029_ASM_C1_D1_1	WC15122_LTS_1_1
WC11029_ASM_C1_D1_2	WC15122_LTS_1_2
WC11125_ASM_E1_E2_H4_1	WC17041_KEITH450___425_1
WC11125_ASM_E1_E2_H4_2	WC17041_KEITH450___425_2
WC12022_AUTO_CL_dot_1	WC17041_KEITH450___425_3
WC12022_AUTO_CL_dot_2	WC20550_CAN_0_52_i_line_1
WC13024_AME_4_5_7_8_1	WC20550_CAN_0_52_i_line_2
WC13024_AME_4_5_7_8_2	

$$
A=\begin{bmatrix}
1 & 0 & 0 & 0 & 0 & 0 & 0 & 0 & 0 & 0 & 0 & 0 & 0 & 0 & 0 & 0 & 0 & 0 & 0\\
0.61 & 1 & 0 & 0 & 0 & 0 & 0 & 0 & 0 & 0 & 0 & 0 & 0 & 0 & 0 & 0 & 0 & 0 & 0\\
0 & 0 & 1 & 0 & 0 & 0 & 0 & 0 & 0 & 0 & 0 & 0 & 0 & 0 & 0 & 0 & 0 & 0 & 0\\
0 & 0 & 0.78 & 1 & 0 & 0 & 0 & 0 & 0 & 0 & 0 & 0 & 0 & 0 & 0 & 0 & 0 & 0 & 0\\
0 & 0 & 0 & 1 & 1 & 0 & 0 & 0 & 0 & 0 & 0 & 0 & 0 & 0 & 0 & 0 & 0 & 0 & 0\\
0 & 0 & 0 & 0 & 0.87 & 1 & 0 & 0 & 0 & 0 & 0 & 0 & 0 & 0 & 0 & 0 & 0 & 0 & 0\\
0 & 0 & 0 & 0 & 0 & 0 & 1 & 0 & 0 & 0 & 0 & 0 & 0 & 0 & 0 & 0 & 0 & 0 & 0\\
0 & 0 & 0 & 0 & 0 & 0 & 0.76 & 1 & 0 & 0 & 0 & 0 & 0 & 0 & 0 & 0 & 0 & 0 & 0\\
0 & 0 & 0 & 0 & 0 & 0 & 0 & 0 & 1 & 0 & 0 & 0 & 0 & 0 & 0 & 0 & 0 & 0 & 0\\
0 & 1 & 0 & 0 & 0 & 0 & 0 & 0 & 0.67 & 1 & 0 & 0 & 0 & 0 & 0 & 0 & 0 & 0 & 0\\
0 & 0 & 0 & 0 & 0 & 0 & 0 & 0 & 0 & 0 & 1 & 0 & 0 & 0 & 0 & 0 & 0 & 0 & 0\\
0 & 0 & 0 & 0 & 0 & 0 & 0 & 0 & 0 & 0 & 0.70 & 1 & 0 & 0 & 0 & 0 & 0 & 0 & 0\\
0 & 0 & 0 & 0 & 0 & 0 & 0 & 0 & 0 & 0 & 0 & 0 & 1 & 0 & 0 & 0 & 0 & 0 & 0\\
0 & 1 & 0 & 0 & 0 & 0 & 1 & 0 & 0 & 0 & 0 & 0 & 0 & 1 & 0 & 0 & 0 & 0 & 0\\
0 & 0 & 0 & 0 & 1 & 1 & 0 & 0 & 0 & 0 & 0 & 0 & 0 & 0 & 1 & 0 & 0 & 0 & 0\\
0 & 0 & 0 & 0 & 1 & 1 & 0 & 0 & 0 & 0 & 0 & 0 & 0 & 0 & 0.89 & 1 & 0 & 0 & 0\\
0 & 0 & 0 & 0 & 0 & 0 & 0 & 0 & 0 & 0 & 0 & 0 & 0 & 0 & 0.57 & 0.79 & 1 & 0 & 0\\
0 & 0 & 0 & 0 & 0 & 0 & 0 & 0 & 0 & 0 & 0 & 0 & 0 & 0 & 0 & 0 & 0 & 1 & 0\\
0 & 0 & 0 & 0 & 0 & 0 & 0 & 0 & 0 & 0 & 0 & 0 & 0 & 0 & 0 & 0 & 0 & 0.78 & 1
\end{bmatrix}
\tag{4.23}
$$

It can be seen from the matrix (4.23) that the utilization ratio of two or three devices in the same device group is very large and has exceeded 0.5.

Conclusion 8: The utilization rate of a device group can be expressed by the average utilization rate of the devices in the device group. Therefore, the bottleneck degree of equipment in the equipment group can be measured by the processing load of the equipment group in the scheduling process.

(2) Correlation analysis of short-term indicators MOV, WIP and TP

The obtained data is consistent with the data obtained from the correlation analysis of production line related performance indicators in Sect. 4.3.2, that is, the correlation coefficients of MOV, WIP and TP are close to 0, and the correlation can be ignored and cannot be replaced by each other.

(3) Correlation analysis of short-term and long-term indicators

(a) Only five dispatching rules are considered

The relationship between MOV and TP, CT and ODR under FIFO, EDD, SPT, CR and LS is as follows:

$$
A=\begin{bmatrix} a_{11} & \cdots & a_{14}\\ \vdots & \ddots & \vdots\\ a_{41} & \cdots & a_{44}\end{bmatrix}=\begin{bmatrix} 1 & 0.51 & 0.55 & -0.40\\ 0.51 & 1 & 0.17 & -0.34\\ 0.55 & 0.17 & 1 & -0.67\\ -0.40 & -0.34 & -0.67 & 1\end{bmatrix}
\tag{4.24}
$$

$$A = \begin{bmatrix} a_{11} & \cdots & a_{14} \\ \vdots & \ddots & \vdots \\ a_{41} & \cdots & a_{44} \end{bmatrix} = \begin{bmatrix} 1 & 0.60 & 0.51 & -0.36 \\ 0.60 & 1 & 0.17 & -0.28 \\ 0.51 & 0.17 & 1 & -0.70 \\ -0.36 & -0.28 & -0.70 & 1 \end{bmatrix} \tag{4.25}$$

$$A = \begin{bmatrix} a_{11} & \cdots & a_{14} \\ \vdots & \ddots & \vdots \\ a_{41} & \cdots & a_{44} \end{bmatrix} = \begin{bmatrix} 1 & 0.54 & 0.56 & -0.41 \\ 0.54 & 1 & 0.17 & -0.29 \\ 0.56 & 0.17 & 1 & -0.71 \\ -0.41 & -0.29 & -0.71 & 1 \end{bmatrix} \tag{4.26}$$

$$A = \begin{bmatrix} a_{11} & \cdots & a_{14} \\ \vdots & \ddots & \vdots \\ a_{41} & \cdots & a_{44} \end{bmatrix} = \begin{bmatrix} 1 & 0.59 & 0.53 & -0.47 \\ 0.59 & 1 & 0.14 & -0.30 \\ 0.53 & 0.14 & 1 & -0.75 \\ -0.47 & -0.30 & -0.75 & 1 \end{bmatrix} \tag{4.27}$$

$$A = \begin{bmatrix} a_{11} & \cdots & a_{14} \\ \vdots & \ddots & \vdots \\ a_{41} & \cdots & a_{44} \end{bmatrix} = \begin{bmatrix} 1 & 0.59 & 0.52 & -0.40 \\ 0.59 & 1 & 0.16 & -0.27 \\ 0.52 & 0.16 & 1 & -0.76 \\ -0.40 & -0.27 & -0.76 & 1 \end{bmatrix} \tag{4.28}$$

It can be seen from the above five matrices that the relationship between MOV and TP is 0.5–0.6, which shows that there is a strong positive correlation between MOV and TP, and the larger MOV is, the greater the output of production line is.

The relationship between MOV and CT is between 0.52 and 0.56. According to common sense, the relationship between them should be negative correlation, but the conclusion here is positive correlation. It can be explained that MOV and CT are not directly related and will be greatly affected by scheduling strategy or WIP level.

The relationship between MOV and ODR is between − 0.36 and − 0.47. According to common sense, the relationship between them should be positive correlation, but the conclusion here is negative correlation. It can be explained that there is no direct correlation between the two, which will be greatly affected by scheduling strategy or WIP level.

The correlation coefficient of CT and ODR between long-term indicators is − 0.67 to − 0.76, showing a strong negative correlation, which indicates that the shorter the processing cycle, the higher the delivery rate.

(b) Consider five working conditions

The relationship among MOV, TP, CT and ODR under five working conditions (WIP quantity is 4000, 5000, 6000, 7000 and 8000 pieces respectively).

$$A = \begin{bmatrix} a_{11} & \cdots & a_{14} \\ \vdots & \ddots & \vdots \\ a_{41} & \cdots & a_{44} \end{bmatrix} = \begin{bmatrix} 1 & 0.05 & 0.68 & -0.39 \\ 0.05 & 1 & 0.23 & -0.42 \\ 0.68 & 0.23 & 1 & -0.81 \\ -0.39 & -0.42 & -0.81 & 1 \end{bmatrix} \tag{4.29}$$

$$A = \begin{bmatrix} a_{11} & \cdots & a_{14} \\ \vdots & \ddots & \vdots \\ a_{41} & \cdots & a_{44} \end{bmatrix} = \begin{bmatrix} 1 & 0.34 & 0.67 & -0.30 \\ 0.34 & 1 & 0.19 & -0.39 \\ 0.67 & 0.19 & 1 & -0.61 \\ -0.30 & -0.39 & -0.61 & 1 \end{bmatrix} \tag{4.30}$$

$$A = \begin{bmatrix} a_{11} & \cdots & a_{14} \\ \vdots & \ddots & \vdots \\ a_{41} & \cdots & a_{44} \end{bmatrix} = \begin{bmatrix} 1 & 0.59 & 0.52 & -0.36 \\ 0.59 & 1 & 0.17 & -0.28 \\ 0.52 & 0.17 & 1 & -0.72 \\ -0.36 & -0.28 & -0.72 & 1 \end{bmatrix} \tag{4.31}$$

$$A = \begin{bmatrix} a_{11} & \cdots & a_{14} \\ \vdots & \ddots & \vdots \\ a_{41} & \cdots & a_{44} \end{bmatrix} = \begin{bmatrix} 1 & 0.64 & 0.51 & -0.47 \\ 0.64 & 1 & 0.15 & -0.28 \\ 0.51 & 0.15 & 1 & -0.72 \\ -0.47 & -0.28 & -0.72 & 1 \end{bmatrix} \tag{4.32}$$

$$A = \begin{bmatrix} a_{11} & \cdots & a_{14} \\ \vdots & \ddots & \vdots \\ a_{41} & \cdots & a_{44} \end{bmatrix} = \begin{bmatrix} 1 & 0.67 & 0.51 & -0.31 \\ 0.67 & 1 & 0.17 & -0.36 \\ 0.51 & 0.17 & 1 & -0.57 \\ -0.31 & -0.36 & -0.57 & 1 \end{bmatrix} \tag{4.33}$$

It can be seen from the above that the correlation coefficient between MOV and TP increases (0.05, 0.34, 0.59, 0.64, 0.67) with the increase of WIP quantity (from 4000 to 8000), indicating that with the increase of WIP, the increase of MOV can directly lead to the increase of TP. The relationship between MOV and TP is greatly affected by working conditions.

With the increase of WIP, the correlation coefficients of MOV and CT are 0.68, 0.66, 0.52, 0.51 and 0.51, respectively, which shows that the increase of WIP quantity leads to more production line blockage, which makes the processing cycle longer and longer, while MOV increases more and more slowly.

Long-term indicators: CT and ODR are negatively correlated. It shows that the decrease of CT can directly cause the increase of ODR.

Conclusion 9: The influence of MOV on TP is reduced under light load, and the influence of MOV on scheduling can be considered under light load, and then the influence of TP on scheduling can be considered under full load and overload.

4.4 Correlation Analysis of Performances Based on Pearson Coefficient

Correlation analysis refers to the analysis of two or more variable elements with correlation, and then evaluates the close degree of correlation between two variable factors. The premise of correlation analysis is that there must be a certain correlation between variable elements. In this section, we will use Pearson correlation coefficient to analyze performance indicators. In statistics, Pearson product moment correlation coefficient is used to measure the correlation between two variables x and y. Among them. Indicates absolute positive correlation and absolute negative correlation. Pearson correlation coefficient between two variables is defined as the ratio of covariance and standard deviation between two variables.

$$\rho_{XY} - 1 \leq \rho_{XY} \leq +1 + 1 - 1$$

$$\rho_{XY} = \frac{E(XY) - E(X)E(Y)}{\sqrt{E(X^2) - E^2(X)}\sqrt{E(Y^2) - E^2(Y)}} \tag{4.34}$$

Using Pearson correlation coefficient formula, the correlation coefficient between short-term performance indexes can be calculated, and many irrelevant variables or variables with little correlation can be eliminated.

In this section, the Pearson correlation coefficient is calculated for the short-term performance index of its production line extracted above and the daily moving steps MOVE, as shown in Fig. 4.12. Because MOVE is an important index to measure the operation performance of semiconductor production line, the higher its value, the higher the processing capacity of semiconductor production line, the higher the utilization rate of equipment, and the higher the number of processing tasks completed by the production line. By summing up the correlation coefficients of MOVE respectively, we can know the influence of a short-term performance index on the total movement and draw a chart of the correlation coefficient changes with time, thus providing help for the follow-up research.

In this section, the correlation coefficients of monthly short-term performance indicators WIP, QL, EQP_UTI and MOVE will be calculated according to the corresponding months, and the law of the corresponding coefficients changing with time will be obtained, which will be plotted and analyzed.

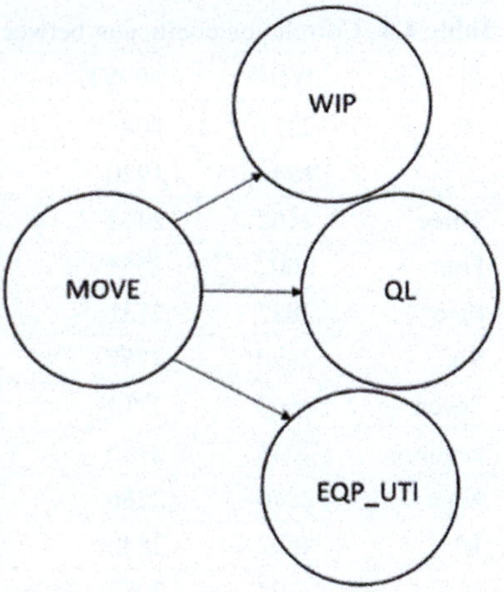

Fig. 4.12 Pearson coefficient diagram of key short-term performance indicators and daily moving steps

4.4.1 Correlation Analysis of Daily WIP and Daily Moving Steps

In this section, the monthly WIP and the daily MOVE in the production line of an enterprise are matched, and their correlation coefficients are calculated by Pearson correlation coefficient formula.

Table 4.4 shows the corresponding average WIP quantity and average daily moving steps in each month, and the correlation coefficient between WIP and MOVE in that month. Figure 4.13 is a trend diagram of the correlation coefficient between WIP and MOVE changing with time. It can be seen from the figure that the correlation coefficient between WIP and MOVE first decreases and then increases with time. The following conclusions can be drawn:

(1) At the beginning and end of the year, the coefficients of WIP and MOVE are close to 1, which means absolute correlation. After analysis, this may be because at the beginning and end of each year, the production line is in the maintenance stage, the actual number of WIP put into the production line is small, the WIP on the production line is basically being processed on equipment, and there are few waiting workpieces in the processing buffer. Therefore, there is an approximate absolute correlation between the daily moving amount MOVE and the current WIP.

(2) In the middle of the year, the production line runs at full capacity and the production capacity is fully utilized. With the increase of orders, many wafers are put into the production line. When the number of wafers to be processed increases, the production line cannot respond in time. There are more factors that influence the daily moving steps MOVE. Therefore, the single variable, WIP, has relatively reduced influence on MOVE.

Table 4.4 Correlation coefficient between WIP and MOVE

Month	WIP	MOVE	Correlation coefficient between WIP and MOVE
One	2211	1093	0.995078
2	1940	1920	0.912262
Three	2462	2474	0.778509
Four	2607	2553	0.439893
Five	2822	2731	0.616548
Six	2860	2757	0.675293
Seven	2660	2714	0.831717
Eight	2045	2190	0.532963
Nine	2276	2266	0.929078
10	4502	2849	0.942395
11	3307	3082	0.731972
12	2484	2499	0.989875

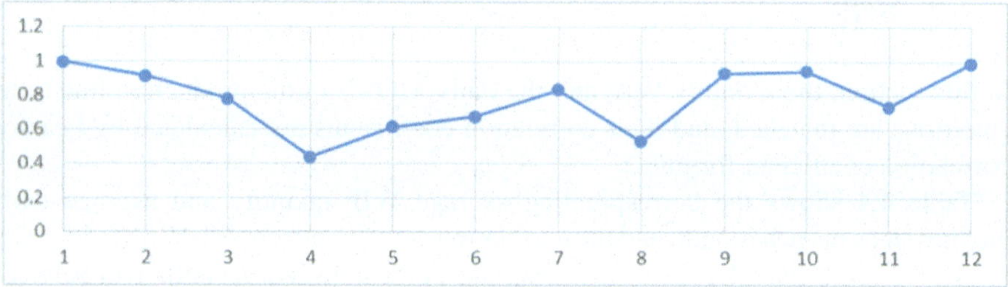

Fig. 4.13 Time variation trend diagram of WIP and MOVE correlation coefficient

When the quantity of WIP is large, the influence of single variable on MOVE is less than that when the quantity of WIP is small.

4.4.2 Correlation Analysis of Daily Queue Leader and Daily Moving Steps

Match the daily queue length QL with the daily moving steps MOVE on the corresponding date and use Pearson correlation coefficient formula to get the correlation coefficient between the monthly queue length QL and MOVE. Table 4.5 shows the monthly average queue length and the corresponding average daily moving steps, and the corresponding correlation coefficients of QL and MOVE in the current month. Figure 4.14 is a trend diagram of correlation coefficient between QL and MOVE changing with time. The following conclusions are drawn from the chart:

Table 4.5 Correlation coefficient table of QL and MOVE

Month	QL	MOVE	Correlation coefficient between QL and MOVE
1	585	1093	0.922042
2	980	1920	0.850559
3	1224	2474	0.498895
4	1336	2553	− 0.13827
5	1416	2731	0.342783
6	1508	2757	0.281906
7	1345	2714	0.628853
8	1001	2190	0.284201
9	1088	2266	0.879445
10	2957	2849	0.877105
11	692	3082	0.500919
12	1248	2499	0.974065

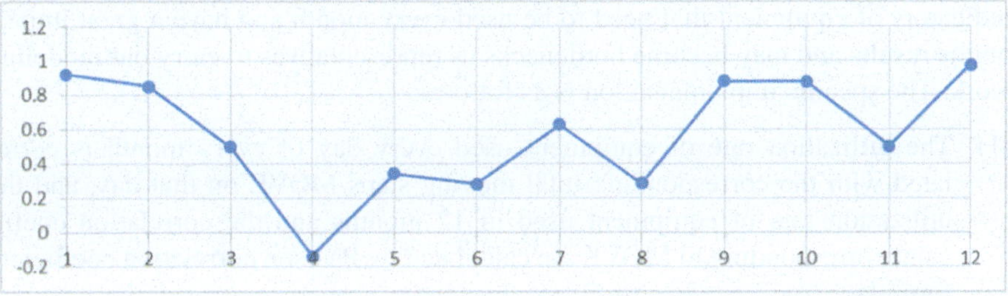

Fig. 4.14 Time variation trend diagram of correlation coefficient between QL and MOVE

(1) The variation trend of the correlation coefficient between queue length QL and total MOVE with time is similar to the curve of the relationship between WIP and total MOVE. At the beginning and end of the year, the correlation coefficient between QL and MOVE is relatively large, and gradually decreases when approaching the middle of the year, and even produces a correlation coefficient close to zero in April. As mentioned above, when the utilization rate of the factory production capacity is large, the factors that have great influence on the total moving amount MOVE increase, so the influence of single variable queue length QL on MOVE is reduced. That is, when the quantity of WIP is large, the influence of single variable on MOVE is less than that when the quantity of WIP is small.

(2) From the comparison of the two correlation coefficient tables, the average correlation coefficient between queue length QL and MOVE is smaller than the average correlation coefficient between WIP and MOVE. Therefore, the influence of queue length QL on total MOVE is less than that of WIP on MOVE.

4.4.3 Correlation Analysis of Daily Equipment Utilization and Daily Moving Steps

Figure 4.15 shows the histogram of the average utilization rate of some equipment in March. Because there are a lot of equipment used every month, and due to the limitation of space, the utilization rate of equipment is not listed in this paper.

It can be seen from Fig. 4.15 that the utilization rate of different equipment varies greatly in the same month, some as high as 70%, some as low as 1%, and the impact of the utilization rate of different equipment on the operation efficiency of the whole factory is obviously different. Therefore, it is inappropriate to use the average utilization rate of all equipment as the short-term performance index of the equipment for establishing the relational model below. However, there are many equipments in the semiconductor production line, and the equipments used in each month are different, so it is inappropriate and undesirable to consider each one, which will increase the complexity of the model and weaken the interpretability of the model. What needs to be established here is the relationship model of macro long-term and short-term performance indicators. Therefore, this paper selects the indicators of equipment that need to be used every month and have a great impact on the results and may become bottlenecks as representatives to carry out modeling work. The specific implementation is as follows:

(1) The utilization rate of equipment used every day of every month is correlated with the corresponding total moving steps MOVE on that day, and the utilization rate of equipment used in 12 months and the correlation coefficient corresponding to MOVE are calculated by Pearson correlation coefficient formula.
(2) Average the correlation coefficients of these devices, sort them from high to low, and select the top 20% devices after sorting, totally 18 devices. The correlation coefficients between these devices and the corresponding daily moving steps are all greater than 0.3.

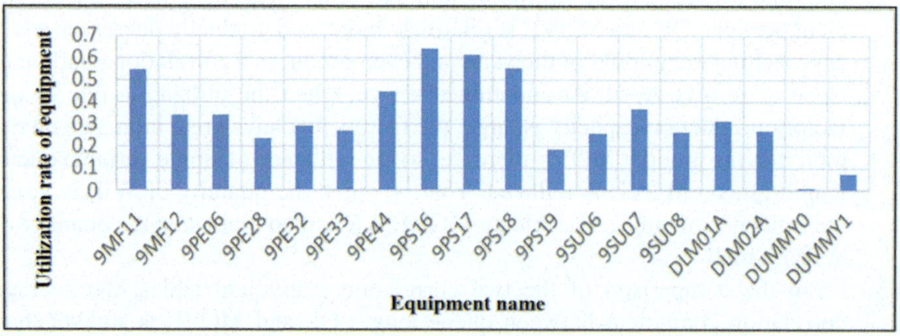

Fig. 4.15 Histogram of utilization rate of some equipment in March

Table 4.6 Correlation coefficient table between equipment utilization rate and MOVE part

Equipment number	Correlation coefficient between EQP_UTI and MOVE of equipment utilization rate
2CL01	0.38009121
7MF04	0.376364211
9CL20	0.368829691
5853	0.363484259
1703	0.350515607
5854	0.349603378
9PS18	0.325888881
5856	0.325549177
5852	0.322465894

In other words, the selected equipment will be used every month, and the usage rate is higher than other equipment, which has a greater impact on the daily moving steps MOVE. In this paper, the corresponding equipment utilization rate of these equipments will be selected as the representative index when establishing the relational model. Table 4.6 is a partial ranking from high to low where the correlation coefficient between the utilization rate EQP_UTI and the total moving amount MOVE of equipment used in 12 months is greater than 0.3.

4.5 Data Set of Performances for Semiconductor Manufacturing System

According to the correlation analysis of the historical data of the actual semiconductor production line, it is not difficult to find that in an actual semiconductor manufacturing system, the correlation coefficient between the long-term and short-term performance indicators fluctuates significantly throughout the year, and the interconnection relationship between the indicators is complex. Based on the aforementioned correlation analysis results and the characteristics of each long-term and short-term performance index, this section selects the short-term performance index with high correlation with the daily moving steps MOVE reflecting the production line efficiency as the model input feature for three predicted performance indexes, i.e., processing cycle, on-time delivery rate and workpiece waiting time, and establishes the corresponding training set and test set.

4.5.1 *Training Set of Processing Cycle and Corresponding Short-Term Performances*

According to the production line characteristics of the actual semiconductor manufacturing system, this section corresponds the long-term performance index processing cycle CT with the screened short-term performance index according to the Lot in Line of the finished workpiece to form a training set. In this way, there are 22 short-term performance indicators as the input of the prediction model.

$$CT_i = \mathcal{F}(WIP_t, \mathrm{QL}_t, MOVE_t, TH_t, EQI\_UTI_{1-18}) \tag{4.35}$$

CT_i indicates THe processing cycle of the workpiece i, t the time when the workpiece enters the production line, WIP_t, QL_t, $MOVE_t$, TH_t: the daily WIP of the whole production line, the daily queue length QL, the daily total moving steps MOVE, and the sunrise piece quantity TH, respectively, and EQI_UTI_{1-18}: the utilization rate of the 18 equipments selected in the correlation analysis above.

Based on the hypothesis of production model under different production conditions, this paper will design and improve the prediction algorithm which accords with the characteristics of production model and establish the relationship model between these 22 short-term performance indexes and processing cycle.

There are 8 kinds of products on the production line of this enterprise, and there are two production lines, namely 5-inch wafer production line and 6-inch wafer production line. There are P1–P3 products for 5-inch line. There are five types of 6-inch lines: P4–P8. Some products have less output and less sample information can be collected. Table 4.7 shows the sample capacity of as-built data that can be collected by each version of wafer. In each modeling, 80% of the data is randomly selected as the training set, and the remaining 20% as the test set.

Table 4.7 Product data capacity of each version

Product type	Data sample capacity (card)
P1	64
P2	43
P3	55
P4	328
P5	121
P6	94
P7	98
P8	123

4.5.2 Training Set of On-Time Delivery Rate and Corresponding Short-Term Performances

Similarly, the training set searches the corresponding short-term performance index for the long-term performance index on-time delivery rate ODR (differentiated product version number) according to the Lot in Line of the finished workpiece. The long-term performance index ODR is generated by rolling cycle. Table 4.8 is a partial training set for on-time delivery rate of product P4. It can be seen from the table that the on-time delivery rate of the production line is generally high and stable, so the influence of equipment utilization rate is not considered for the on-time delivery rate data set, that is, only the relationship between the on-time delivery rate ODR and the daily WIP quantity, sunrise piece quantity TH, daily moving steps MOVE and daily queue length QL in short-term performance indicators is considered.

$$ODR_{t,i} = \mathcal{F}(WIP_t, \mathrm{QL}_t, MOVE_t, TH_t) \tag{4.36}$$

$ODR_{t,i}$ indicates the on-time delivery rate of a product type to which the workpiece i belongs within the average processing cycle T from the day t to the day t + T. It indicates a certain time and day when the card workpiece enters and leaves the production line, and WIP_t, QL_t, $MOVE_t$, TH_t respectively indicates the daily WIP, the daily queue length QL, the daily moving steps MOVE and the sunrise piece quantity th of the whole production line.

Table 4.8 P4 partial training set for forecasting on-time delivery rate

WIP work in process	Queue leader QL	Total MOVEment amount move	Chip output TH	On-time delivery rate ODR
1628	1519	853	136	0.933333
1677	1565	878	142	0.833333
1754	1638	919	142	0.933333
1822	1715	951	143	0.933333
1883	1779	981	142	0.933333
1888	1786	985	144	0.966667
1866	1761	975	140	0.933333
1866	1761	975	140	0.933333

4.5.3 Training Set of Waiting Time and Corresponding Short-Term Performances

According to the card number ID of the completed workpiece, the training set links the short-term performance index Waiting Time (distinguishing product version number) with the other 22 short-term performance indexes mentioned above, including the daily WIP, sunrise piece quantity th, daily moving steps MOVE and daily queue length QL of the whole production line with the average equipment utilization rate of 18 typical equipment.

$$WT_i = \mathcal{F}(WIP_t, \mathrm{QL}_t, MOVE_t, TH_t, EQI\_UTI_{1-18}) \tag{4.37}$$

WT_i indicates the time and time of the workpiece i waiting for processing in the buffer during the whole life cycle of the production line. T indicates THe time when the card workpiece enters the production line, WIP_t, QL_t, $MOVE_t$, TH_t respectively indicates the daily WIP, the daily queue length QL, the daily moving steps MOVE, and the sunrise piece quantity th of the whole production line on the day when a card workpiece enters the production line, and EQI_UTI_{1-18} indicates the utilization rate of 18 equipments selected in the correlation analysis.

4.6 Summary

This chapter introduces the characteristics of long-term and short-term performance indexes of semiconductor production line in detail. Using the historical production data of an actual semiconductor enterprise, the correlation analysis results are discussed according to the key long-term and short-term performance indexes and the actual production scenarios of semiconductors. According to the results of correlation analysis, a reasonable training set and test set are established for the three performance indicators as prediction targets, which will prepare data for the establishment of prediction models for different production conditions and different performance indicators.

References

1. Pan CR, Zhou MC, Qiao Y, et al (2017) Scheduling cluster tools in semiconductor manufacturing: recent advances and challenges. IEEE Trans Autom Sci Eng
2. Wang Z, Wu Q (2002) Research on control and scheduling of semiconductor production line. Comput Integr Manuf Syst 8(8):607–611
3. Cao G, You H, Jiang Z, et al (2008) Dynamic hierarchical planning and scheduling method of semiconductor production line based on TOC. Mod Mach Tool Autom Process Technol 2008(10).

4. Shi B, Qiao F, Ma Y (2009) Research on dynamic scheduling of semiconductor production line based on fuzzy Petri net reasoning. Mechatronics 15(4):29–32
5. Wang L, Lu X, Agent Y (2007) Research on dynamic scheduling of semiconductor production line based on multi-agent technology. Comput Eng 33(13):4–6
6. Wu Q, Ma Y, Li L et al (2015) Dynamic scheduling method of semiconductor production line driven by data. Control Theory Appl 32(9):1233–1239
7. Qiao F, Xu X, Fang M et al (2007) Study on the performance index system of semiconductor wafer production line scheduling. J Tongji Univ: Nat Sci Ed 35(4):537–542
8. Mönch L, Fowler JW, Dauzère-Pérès S et al (2011) A survey of problems, solution techniques, and future challenges in scheduling semiconductor manufacturing operations. J Schedul 14(6):583–599
9. Ma Y, Qiao F, Chen X et al (2015) Dynamic scheduling method of semiconductor production line based on support vector machine. Comput Integr Manuf Syst 21(3):733–739
10. Lee YF, Jiang ZB (2009) Liu HR (2009) Multiple-objective scheduling and real-time dispatching for the semiconductor manufacturing system. Comput Oper Res 3:866–884
11. Baez Senties O, Azzaro-Pantel C, Pibouleau L (2009) A neural network and a genetic algorithm for multiobjective scheduling of semiconductor manufacturing plants. Ind Eng Chem Res 2009(21):9546–9555
12. Tang J, Wang X, Kaku I et al (2009) Optimization of parts scheduling in multiple cells considering intercell move using scatter search approach. J Intell Manuf 21(4):525–537
13. Li D, Meng X, Li M et al (2016) An ACO-based intercell scheduling approach for job shop cells with multiple single processing machines and one batch processing machine. J Intell Manuf 27(2):283–296
14. Elmi A, Solimanpur M, Topaloglu S et al (2011) A simulated annealing algorithm for the job shop cell scheduling problem with intercellular moves and reentrant parts. Comput Ind Eng 61(1):171–178
15. Jia P, Wu Q, Li L (2014) Dynamic dispatching method of semiconductor production line driven by performance index. Comput Integr Manuf Syst 20(11).
16. Su G, Wang X (2011) Establishment of improved petri net model and optimal scheduling of semiconductor manufacturing system. Syst Eng Theory Pract 31(7):1372–1377
17. Zhang H, Jiang Z, Guo C, Liu H (2006) Real-time scheduling simulation platform of wafer manufacturing system based on EOPN. J Shanghai Jiaotong Univ 40(11):1857–1863

Chapter 5
Data-Driven Release Control of Semiconductor Manufacturing System

Semiconductor manufacturing system is known as the most complex manufacturing system because of its complex reentrant process flow, mixed processing mode, high uncertainty and rapid product and technology update. At present, the scheduling problem of semiconductor wafer manufacturing system can be generally divided into six aspects: releasing strategy, dispatching rules, batch processing equipment scheduling, bottleneck equipment scheduling, equipment maintenance scheduling and rescheduling. Because the work-in-process level of the early wafer factories was high, and the dispatching rules were easy to apply and made the work-in-process level significantly improved, the enterprises had already recognized it highly. Therefore, dispatching rules became the focus of research in the following decades, ignoring the research on the releasing strategy. However, with the maturity of the research on dispatching rules, the improvement of work-in-process level has encountered bottlenecks, and people have come back to re-examine the research on the releasing strategy. Accordingly, the research on the releasing strategy has a rising trend in recent years.

5.1 Common Release Strategies of Semiconductor Manufacturing Systems

Release control is an important part of the scheduling of semiconductor manufacturing system, which is at the front end of the scheduling system of semiconductor manufacturing system and plays an important role in the scheduling of the whole semiconductor manufacturing process. Release control affects other types of scheduling and is of great significance to improve the overall performance of semiconductor manufacturing system [1–3].

L. Li et al., *Data-Driven Scheduling of Semiconductor Manufacturing Systems*, Advanced and Intelligent Manufacturing in China,
https://doi.org/10.1007/978-981-19-7588-2_5

Because the semiconductor manufacturing system has the following characteristics:

1. Silicon wafer processing has almost harsh requirements on the environment. The whole process should be carried out under the condition of "clean room". The longer it is exposed to the air, the greater the possibility of pollution [4];
2. Silicon wafers are limited by the "maximum waiting time limit". If the waiting time in front of the equipment exceeds the "time limit", the device will fail and be redone [5];
3. For order-oriented multi-variety, small-batch, resource-limited silicon wafer processing lines with "reflow", it is necessary to meet various customer requirements for delivery time [6].

Therefore, too much releasing and too high WIP will reduce the qualified rate of silicon wafers on the one hand and prolong the manufacturing cycle on the other hand; However, if the releasing quantity is too small, some equipment will be idle, resulting in the waste of system resources [7]. The releasing strategy has an important influence on the performance of semiconductor manufacturing system.

Release control determines the type, quantity and releasing time of products put into the production line, to give full play to the production capacity of the production system as much as possible [8]. The purpose of release control is to make the utilization rate of key equipment in the whole production line at a higher level, at the same time increase the output per unit time of the production line as much as possible and reduce the processing cycle of products to enhance the efficiency of enterprises [9]. Therefore, release control plays an important role in the whole multi-input production process scheduling, which is of great significance to improve the overall performance of complex multi-input production systems.

At present, the research on the releasing strategy of multi-input production system at home and abroad mainly focuses on two directions: common release control and improved release control [10].

5.1.1 Common Release Control

The current complex multi-input production system mainly adopts commonly used release control methods, which can be divided into static release control strategy and dynamic release control strategy as shown in Fig. 5.1.

5.1.1.1 Static Release Control Strategy

Static release control strategy is to feed without considering the real-time information feedback of production line, which belongs to open loop releasing strategy. Specifically, it can be divided into two types: releasing based on time interval and releasing based on releasing list [11].

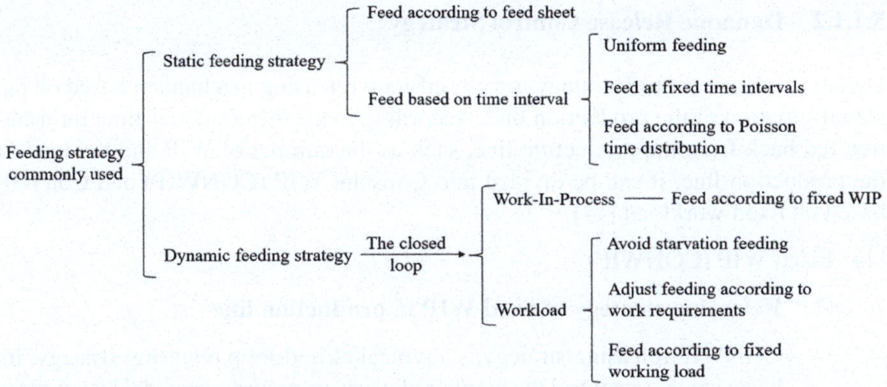

Fig. 5.1 Classification of releasing strategies

Releasing based on the releasing list is to set the priority order of orders according to the urgency of delivery date, set the releasing time point of each order, and put the corresponding orders into the production line according to the releasing time point [12].

The releasing strategy based on time interval includes Constant Time (CONTime), poisson time distribution (EXPTime) and Uniform Release (UNIF).

CONTime keeps the releasing time interval on the production line unchanged, that is, releasing follows Formula (5.1):

$$\mathrm{T} \leq |24/R_d| \tag{5.1}$$

In Formula (5.1), T represents releasing interval time, the unit is hours; R_d represents the total number of releasing workpieces determined every day.

The time interval of EXPTime is obtained according to Poisson distribution.

UNIF refers to putting all the workpieces to be fed into the production line in a certain order at a certain moment, without considering releasing into the production line in batches. This releasing method needs to set the priority of releasing workpieces based on fully analyzing the performance of the production line and decide the releasing sequence according to this priority. It has the advantages of simplicity and rapidity and does not need to adjust the releasing decision according to the situation on the production line, thus simplifying the releasing mechanism and improving the response speed. Its disadvantage is that there is no scientific and accurate mechanism to set the releasing sequence, but only based on experience [13].

The releasing strategy based on time interval has certain advantages, that is, it is simple and easy to realize. Its disadvantage is that it does not consider the actual situation of the production line, which easily leads to the backlog of workpieces.

5.1.1.2 Dynamic Release Control Strategy

Dynamic release control strategy is a closed-loop releasing mechanism based on the actual situation of the production line. According to the different real-time information fed back from the production line, such as the number of WIP and the load of the production line, it can be divided into Constant WIP (CONWIP) and ConWIP based on fixed workload [14].

(1) Fixed WIP (CONWIP)

(1) **Releasing strategy of fixed WIP in production line**

CONWIP releasing strategy is a typical closed-loop releasing strategy. Its basic idea is to control the number of work-in-process at an ideal level as far as possible, that is, when a workpiece leaves the system, a new workpiece of the same kind has the right to enter the system, thus keeping the number of work-in-process on the production line unchanged. CONWIP strategy can effectively control inventory and production, thus controlling the whole production [15]. In CONWIP releasing strategy, how to determine the ideal WIP quantity is the key. If the ideal WIP is too high, the production line can not be effectively controlled; If the ideal WIP is too low, it may reduce the production capacity of the production line.

The formula (5.2) is commonly used to determine the ideal WIP quantity under CONWIP releasing strategy:

$$\mathrm{TH}(\omega) = \omega r_b / (\omega + W_0 - 1) \tag{5.2}$$

In which, ω represents the expected target WIP quantity of chip manufacturing line. $\mathrm{TH}(\omega)$ represents the desired productivity under the target WIP quantity. r_b represents the processing rate of the processing center where the bottleneck equipment is located, that is, the number of cards processed in the processing center of the bottleneck equipment per time unit. W_0 represents the processing capacity of the processing center where the bottleneck equipment is located T_0 is the sum of the average processing time of the processing center , $W_0 = r_b T_0$;

Example 5.1 It is known that the production line has four machining centers, namely lithography, etching, oxidation and implantation. The technical parameters of each machining center are shown in Table 5.1. The ideal productivity is 0.33 cal/h, and the most ideal WIP level of the production line is obtained.

Solution: According to the data in Table 5.1, the ideal productivity is 0.33 cal/h, which can be obtained TH(ω)

$$r_b = 1/[(2 + 2.5)/2] = 0.44\,\text{Cards/h}$$

$$T_0 = (2 + 2.5)/2 + (1.1 + 1.3)/2 + (1.2 + 1.4 + 1.5)/3 + (0.3 + 0.4)/2 = 6.3\text{h}$$

Table 5.1 Technical parameters of machining center in model

Machining center	State	Equipment	Program	Processing time (h)	Program	Processing time (h)	Program	Processing time (h)
Photoetching	Bottleneck	STP	M1	2	M2	2.5		
Corrosion	Non-bottleneck	EH1	M1	1	M2	1.2		
		EH2	M1	1.1	M2	1.3		
Oxidation	Non-bottleneck	BTU	M1	1.2	M2	1.4	M3	1.5
Thin film	Non-bottleneck	IMP	M1	0.3	M2	0.4		

$$W_0 = r_b T_0 = 0.44 \times 6.3 = 3\,\text{Cards}$$

It can be obtained from Formula (5.2):

$$\omega^* \text{ TH}(\omega) + W_0 * \text{TH}(\omega) - \text{TH}(\omega) = \omega r_b$$

$\omega = \text{TH}(\omega)(1 - W_0)/(\text{TH}(\omega) - r_b) = 0.33 * (1 - 3)/(0.33 - 0.44) = 6\,\text{Cards}$
That is, the ideal WIP level on the production line is 6 cards.

There are three limitations of CONWIP release control: first, it only macroscopically controls the WIP quantity of the whole production line. however, there are many processing zones in a production line, and only paying attention to the WIP quantity of the whole production line without controlling the WIP quantity of each processing zone, it is likely that there will be too many WIP quantities in some processing zones and too few WIP quantities in some processing zones, which is unfavorable for the operation of the whole production line; Second, when there are many product versions on the production line, especially when the processing flow is very different, it is difficult to determine the appropriate WIP quantity according to the formula; Third, because only the quantity of WIP is controlled, and the processing degree of WIP is not considered, in extreme cases, most of the workpieces may be in the processing stage of the first circuit or in the processing stage of the last circuit. Therefore, CONWIP releasing strategy can not fully meet the requirements of production line control and dynamic change of orders.

(2) **Layered fixed in-process releasing**

In order to overcome the deficiency that CONWIP releasing strategy can't control the processing degree of WIP, a Layerwise CONWIP (LCONWIP) releasing strategy was proposed. The idea of this releasing strategy comes from the characteristics of semiconductor manufacturing process [16]. In the process of semiconductor manufacturing, semiconductor components are of hierarchical structure, and each layer is produced in a similar way (see Sect. 2.3 for details). Therefore, LCONWIP will set an ideal value for the number of work-in-process processed at each level of the production line.

In the simplest case, the ideal WIP quantity of the production line can be determined according to the Formula (5.2), and then the ideal WIP value can be evenly distributed to each layer of WIP according to the processing level of the workpiece, as the ideal WIP value of each layer. As shown in Formula (5.3):

$$\omega_i = \omega / n \tag{5.3}$$

where ω_i is the ideal work-in-process value of the i_th layer, i = 1, ⋯ , n, n is the total number of layers of the processed semiconductor products, and is generally counted as the total lithography times; ω Is that ideal WIP value of the whole production line.

Table 5.2 Product processing flow

Product version	Working procedure	Equipment (program)
Version 1 (B1)	S1	STP (M1)
	S2	EH1 (M1)、EH2 (M1)
	S3	BTU (M1)
	S4	STP (M2)
	S5	IMP (M1)
	S6	BTU (M3)
	S7	STP (M2)
	S8	EH1 (M2)、EH2 (M2)
Version 2 (B2)	S1	STP (M1)
	S2	EH1 (M1)、EH2 (M1)
	S3	BTU (M3)
	S4	STP (M2)
	S5	IMP (M2)
	S6	EH1 (M2)、EH2 (M2)

Taking Example 5.1 as an example, assuming that two product versions are produced on the production line, and the processing flow of each product version is shown in Table 5.2, the processing layer number of product version 1 is three, and the processing layer number of product version 2 is two. As can be seen from Example 5.1, the best WIP value of this production line is 6 cards, which are evenly distributed to two product versions, that is, the WIP target value of each product version is 3 cards. Using Formula (5.2), the target WIP value of each layer can be further determined, that is, the target WIP value of the first layer, the second layer and the third layer of product version 1 is 1 card respectively; However, the first layer of product version 2 is 1 card, and the second layer is 2 cards (when decimal numbers appear, round up and put the larger value on the last layer).

(2) Releasing strategy based on workload

Workload-based releasing strategies can be specifically divided into three releasing strategies: Constant Load (CONLOAD), Starvation Avoidance (SA) and Workload Regulation (WR).

(1) **Fixed workload releasing strategy (CONLOAD)**

Unlike CONWIP and LCONWIP releasing strategies, which only focus on the number of works in process, CONLOAD releasing strategy pays more attention to the workload of bottleneck equipment, which is specifically defined as a release control method that keeps the workload of bottleneck equipment at a certain fixed value. For example, the target load of bottleneck equipment can be set to 90% of its daily processing capacity.

In the semiconductor manufacturing production line, the investment of each card of new workpiece will increase the total load of the production line, and at the same time, the workload of bottleneck equipment will increase correspondingly [17]. Therefore, only when the total man-hours required for queuing all the workpieces in front of the bottleneck equipment of the production line is lower than the target load of the bottleneck equipment, can new workpieces be put in.

The advantage of CONLOAD releasing strategy is that its parameter setting is more intuitive than the traditional CONWIP releasing strategy, and it can adapt to the change of product variety.

(2) **Releasing strategy to avoid hunger (SA)**

The connotation of releasing strategy to avoid hunger is to improve the utilization rate of bottleneck equipment as much as possible while controlling the WIP level. Its core idea is very simple: to reduce inventory, you don't want to put in new workpieces, but the end result is that bottleneck equipment is hungry and there are no finished workpieces [18]. Therefore, new workpieces must be put into the production line in time to avoid idle bottleneck equipment due to lack of workpieces.

The queuing problem of bottleneck equipment is similar to the inventory problem. In inventory control, the main goal is to achieve the compromise between inventory cost and shortage cost. If the customer demand and the order lead time (that is, the delay time between the order ordering time and the order delivery time) are determined, the inventory control is very simple [19]. However, in practice, demand and order lead time are uncertain. To meet customers' needs in time, the concept of safety stock is put forward in inventory control. In the reorder point inventory control system, if the inventory is reduced below the safety stock, a new order will be placed to ensure the safety stock. If the safety stock is large enough, it will ensure that the new orders will arrive in time with sufficient probability to avoid the shortage of stock.

Assuming that T_E and T_R are the time to deplete inventory and replenish inventory, the expected time to deplete inventory $E(T_E)$ can be obtained by the ratio of actual inventory I to demand rate $\overline{d}$, as shown in Formula (5.4):

$$E(T_E) = I/\overline{d} \tag{5.4}$$

If $T_E < T_R$, there will be a shortage of goods. To avoid shortage, new orders should be placed when the inventory meets the following conditions, as shown in Formula (5.5):

$$I < \overline{d} \times T_R + ss \tag{5.5}$$

In which: ss refers to safety stock.

Similarly, in the semiconductor production line, the change of inventory of bottleneck equipment comes from uncertain demand (equipment failure and repair time are uncertain), uncertain lead time for new input workpieces to reach bottleneck equipment and WIP from other work centers. In SA, firstly, the concept of virtual inventory is defined, which is used to express the total working hours of workpieces waiting to be processed in front of bottleneck equipment and workpieces waiting to be processed in a given time. The given time is generally expressed by the estimated time when the newly input workpiece reaches the bottleneck equipment for the first time. In addition, the number of workpieces that can be completed within the expected repair time of the equipment currently in repair state at the bottleneck work center is also a part of the virtual inventory. Like inventory control, when the virtual inventory falls below the predetermined level, it is necessary to put new parts into the production line.

Obviously, the goal of SA is to make the newly released workpieces reach the bottleneck equipment in time to avoid starvation of the bottleneck equipment. Safe virtual inventory is the control parameter of the system. Increasing the safety virtual inventory will increase the average inventory level, but it can reduce the idle probability of bottleneck equipment due to lack of workpieces and increase the output.

Based on the above analysis, the formal definition of hunger avoidance releasing strategy (SA) in single product production line with only one bottleneck work center is given below. Let the bottleneck work center be B, where there is the same equipment m, and its average repair time is $MTTR_B$.

Let the number of workpieces (including the workpieces being processed and the workpieces waiting in line for processing) in the current process i beK_i. The i-th process is completed by using the machining center w_i, and the machining time isd_i. Set i_0 as the number of processes visited B for the first time, as shown in Formula (5.6).

$$i_0 = \min\{\mathrm{i}|w_i = B\} \tag{5.6}$$

Set S_B as the set of processes that the bottleneck work center can complete, as shown in Formula (5.7):

$$S_B = \{\mathrm{i}|w_i = B\} \tag{5.7}$$

Set FB as the set of all processes before the first visit, as shown in Formula (5.8):

$$\mathrm{F} = (1, \cdots, \mathrm{i}_0 - 1) \tag{5.8}$$

The definition of L is the total processing time of the operation from the first operation to the first visit B, as shown in Formula (5.9):

$$L = \sum_{i=1}^{i_0-1} d_i \tag{5.9}$$

Let n_i be the number of processes visited B by the workpiece whose current process is i the next time, and then define P as the set of processes whose processing time is less than L the sum of the processes before the next visit B, as shown in Formula (5.10):

$$P = \left\{ i \middle| \sum_{j=1}^{n_i-1} d_j < L \right\} \tag{5.10}$$

Let $Q = F \cup P \cup S_B N(B)$ be a collection of key processes, and then define the number of equipment being repaired in the bottleneck work center. The estimated total equipment repair time is shown in Formula (5.11):

$$R = MTTR_B \times N(B) \tag{5.11}$$

The virtual inventory definition of bottleneck work center W is shown in Formula (5.12):

$$W = \left(R + \sum_{i \in Q} K_i d_{n_i} \right) / m \tag{5.12}$$

If the virtual inventory of the bottleneck work center is less than αL, as shown in Formula (5.13):

$$W < \alpha L \ (\alpha > 0) \tag{5.13}$$

α represents the satisfaction coefficient which can be set artificially, so the bottleneck equipment is in danger of starvation, so new workpieces are required to be put into the production line.

The above definition is the releasing strategy of hunger avoidance when there is only one fixed bottleneck work center on the production line. In the actual semiconductor manufacturing environment, the position of bottleneck work centers may drift with the different varieties of products flowing online at the same time. At this time, the idea of SA can be extended to the environment with multiple bottlenecks, that is, if the queuing level of all work centers falls below the level of safe virtual inventory, releasing will be carried out.

(3) **Workload adjustment releasing strategy**

The core idea of workload adjustment releasing strategy is to adjust the workload of each processing area of semiconductor manufacturing line by releasing, so as to achieve the best performance [20].

Workload adjustment is generally linked with the adjustment of production capacity. When the workload changes, the production capacity may also change [21]. The goal of workload adjustment releasing strategy is to increase the production capacity of the whole production line through workload adjustment.

Workload adjustment includes three aspects: workload description, workload prediction and workload control. The workload description needs to consider the measurement and modeling of production load; Workload forecasting refers to predicting the future use of resources through observation and measurement of data; Workload control refers to considering the forecast demand into the capacity plan, continuously monitoring the operation of the workload, and making possible load adjustment.

For a specific semiconductor manufacturing line, some processing areas are in a bottleneck state every day. If a certain processing area is in a bottleneck state, there must be many workpieces stuck in front of this processing area waiting for processing, and the biggest factor affecting the processing cycle of workpieces is the waiting time of workpieces [22]. For the whole production line, there are more workpieces waiting to be processed in some processing areas, so other processing areas may be idle, and the equipment in these idle processing areas cannot be fully utilized. By adjusting the workload of each processing area through releasing, we can expect to obtain optimized semiconductor manufacturing performance, which is the goal of the workload adjustment strategy.

5.1.2 Improved Release Control Strategy

There are two main ways to study the improved release control strategy: (1) comprehensive releasing strategy; (2) The releasing strategy of sub-product hierarchical control. The comprehensive releasing strategy contains two ideas: one is the integration of many common releasing strategies; Second, the combination or integration of releasing and dispatching [23]. The releasing strategy of sub-product hierarchical control is put forward because it considers that a variety of products are mixed in the same production line and are being processed at the same time. For example, common release control strategies such as CONWIP are adopted, which do not consider the types of products and the processing progress, resulting in excessive concentration of WIP in a certain processing area and affecting the output rate of the production line [24].

Based on CONWIP and CONLOAD, Qi et al. [25–27] put forward a new dynamic releasing strategy-WIPLctrl, which overcomes the shortcoming that CONLOAD only considers the load of bottleneck equipment. WIPLCtrl release control strategy takes the total remaining processing time (WIPLoad) of all workpieces on the production line as a measurement index, monitors the value of WIPLoad on the production line in real time through a closed-loop control system, and adjusts the releasing strategy accordingly. In this paper, WIPLCtrl release control strategy is applied to the actual semiconductor multiple-entry production line, which is compared with WR, CONWIP and UNIF according to the output of the production line. When the output of the production line is low, the average processing cycle obtained by WIPLCtrl release control strategy is the smallest; When the output of the production line is relatively high, the WIPLCtrl release control strategy can obtain a smaller average processing cycle, and its advantages are more obvious than the other three releasing strategies. Moreover, with the increase of disturbance in production line, the reliability and robustness of WIPLCtrl release control strategy is also the highest.

Bahaji et al. [28] studied various combinations of CONWIP, push release control strategy and common dispatching rules through simulation. The simulation results pointed out the advantages and disadvantages of each combination and described its application scope in detail. Wang et al. [29] put forward a Compound Priority Dispatching rule (CPD), which takes work-in-process management and release control into account while dispatching work, integrates the current initial state of production line, the number of work-in-process and upstream and downstream process information, and puts forward a formula for calculating the compound priority of workpieces. Simulation results show that compared with FIFO and SRPT dispatching rules, CPD can significantly reduce the Mean Total Queue Time (MTQT) of production line and increase the production rate of production line. Li et al. [30] proposed a Monte Carlo simulation method based on meta-model to capture the dynamic and random behavior in semiconductor manufacturing system and optimize the releasing plan in real time according to the performance index. Simulation results show that this method can effectively improve the releasing plan and improve the performance of the production line. Chen et al. [31] proposed a release control strategy based on extreme learning machine, aiming at the fact that the threshold of dynamic release control strategy is often determined according to trial and error method, but not dynamically adjusted according to real-time state information. It mainly establishes the learning mechanism of workpiece information, real-time state of production line and release control strategy. In actual production, the releasing sequence and timing are determined according to the established learning mechanism. Simulation results show that this strategy can improve two performance indexes, namely, chip yield and processing cycle. However, it can't cope with the drift of bottleneck equipment when the product types and proportions change.

Ya et al. [32] put forward the concept of integrating dispatching rules with releasing policies (DW&SCP) for the first time. In terms of dispatching rules, DW&SCP divides the equipment on the production line into three categories according to the equipment utilization rate and entropy value representing dispatching complexity: non-bottleneck equipment, bottleneck equipment with smaller entropy value and bottleneck equipment with larger entropy value. These three types of equipment adopt different combinations of WIP control and output control as their dispatching rules. In the aspect of release control strategy, DW&SCP

adopts CONWIP release control in the early stage and adopts fixed time interval releasing method with time interval as production rate in the later stage, thus balancing the contradiction between WIP quantity and releasing rate. The result of simulation shows that DW&SCP is superior to Fewest Lots in the Next Queue (FLNQ), Shortest Process Time, Any combination of dispatching rules such as SPT), Least Amount of Expected Workper Machine (LWNQ/M) and UNIF, EXPTime and CONWIP release control strategies.

Sun et al. [33] first put forward the release control strategy of Dynamic Classified WIP (DC-WIP) based on the concept of sub-orders. Sub-orders refer to dividing a customer's order into multiple sub-orders according to product types, so that each sub-order only includes one type of product. At the same time, all sub-orders from the same order have the same attributes in customer importance and order urgency, but they are different in order size, processing cycle and profit expectation. Therefore, the author uses the priority ranking method based on fuzzy algorithm to get the priority of each sub-order. At last, the author obtains the optimal WIP value (WIP_i) of each sub-order through TOC and Little formula. The releasing idea of DC-WIP is to judge whether the WIP quantity of the current sub-order on the production line $WIP_{(avr)i}$ is satisfied $WIP_{(avr)i} = WIP_i$ or not according to the priority of the sub-order, and if so, put the products of the sub-order until $WIP_{(avr)i} = WIP_i$ is satisfied. In this paper, the author applies the DC-WIP release control strategy to the mini-fab model. By comparing this releasing strategy with CONTime and AVR-WIP (that is, WIP_i in every sub-order is equal), the DC-WIP release control strategy is superior to the other two releasing strategies in terms of on-time delivery rate (OTD) and average processing cycle.

5.1.3 Research Status of Release Control Strategies

Through the above research, it is found that the commonly used release control strategy and the improved release control strategy have certain limitations.

Among the commonly used release control strategies, the static release control strategy often does not consider the real-time status information on the production line, while the dynamic release control strategy considers only one aspect of the real-time status information, such as CONWIP only considers the work-in-process quantity of the production line, WR only considers the workload on bottleneck equipment. In addition, the threshold value in the dynamic release control strategy is often determined according to trial and error method, and is not dynamically adjusted according to real-time state information, so the static releasing method and dynamic releasing method in the commonly used release control strategy do not fully consider the real-time state of the production line, which is one-sided.

The improved release control strategy integrates the advantages of various release control strategies, but the effect and practicality are contradictory, that is, the effect and practicality of releasing strategy are often negatively correlated. The effect of this releasing method may be remarkable, but its complex decision-making mechanism often needs a lot of calculation time, which affects the efficiency of releasing decision-making.

Therefore, it is necessary to study a release control strategy which comprehensively considers the releasing order information and the real-time status information of the production line and has optimization capability.

5.2 Release Control Strategy Based on Extreme Learning Machine

Extreme Learning Machine, (ELM) is a new type of feedforward neural network learning method developed in recent years, which is widely used, such as building a model to predict the elasticity of cohesive soil, fault diagnosis of rotating machinery, face recognition, motion recognition and so on. ELM has the advantages of fast training speed, global optimal solution, and good generalization performance, so ELM is adopted as the learning mechanism of release control strategy in this section.

The typical structure of ELM is shown in Fig. 5.2. It consists of input layer nodes, hidden layer nodes and output layer nodes. Among them, the input layer weight represents the gain between the input node and the hidden layer node, as shown in Formula (5.14):

$$\omega_{n\times l} = \begin{bmatrix} \omega_{11} & \omega_{12} & \cdots & \omega_{1l} \\ \omega_{21} & \omega_{22} & \cdots & \omega_{2l} \\ \vdots & \vdots & \ddots & \vdots \\ \omega_{n1} & \omega_{n2} & \cdots & \omega_{nl} \end{bmatrix} \tag{5.14}$$

Fig. 5.2 Structure diagram of extreme learning machine

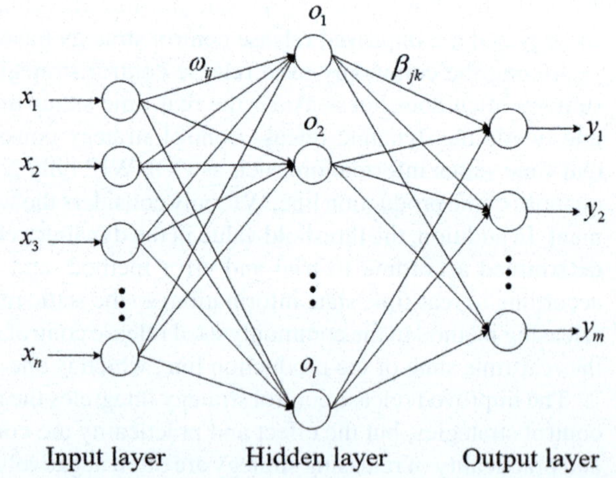

In Formula (5.15), the output layer weight β_{jk} represents the gain between the hidden layer j-th neuron and the k-th output.

$$\beta_{l\times m} = \begin{bmatrix} \beta_{11} & \beta_{12} & \cdots & \beta_{1m} \\ \beta_{21} & \beta_{22} & \cdots & \beta_{2m} \\ \vdots & \vdots & \ddots & \vdots \\ \beta_{l1} & \beta_{l2} & \cdots & \beta_{lm} \end{bmatrix} \tag{5.15}$$

$b_{l\times 1}$ said the threshold of hidden layer, as shown in Formula (5.16):

$$b_{l\times 1} = \begin{bmatrix} b_1 & b_2 & \cdots & b_l \end{bmatrix}' \tag{5.16}$$

ELM algorithm randomly initializes the input layer weight matrix $\omega_{n\times l}$ and the hidden layer threshold matrix $b_{l\times 1}$, without iterative training and learning data. Let the training set have Q samples, the input matrix is X and the output matrix is Y which can be expressed by the following Formulas (5.17) and (5.18) respectively:

$$X_{n\times Q} = \begin{bmatrix} x_{11} & x_{12} & \cdots & x_{1Q} \\ x_{21} & x_{22} & \cdots & x_{2Q} \\ \vdots & \vdots & \ddots & \vdots \\ x_{n1} & x_{n2} & \cdots & x_{nQ} \end{bmatrix} \tag{5.17}$$

$$Y_{m\times Q} = \begin{bmatrix} y_{11} & y_{12} & \cdots & y_{1Q} \\ y_{21} & y_{22} & \cdots & y_{2Q} \\ \vdots & \vdots & \ddots & \vdots \\ y_{m1} & y_{m1} & \cdots & y_{mQ} \end{bmatrix} \tag{5.18}$$

Assuming that the activation function of the hidden layer is $g(x)$, and the Sigmoid function is generally taken, it can be known that the output of the network is shown in Formulas (5.19) and (5.20).

$$T_{m\times Q} = [t_1, t_2, \ldots, t_Q] \tag{5.19}$$

$$t_j = \begin{bmatrix} t_{1j} \\ t_{2j} \\ \vdots \\ t_{mj} \end{bmatrix} = \begin{bmatrix} \sum_{i=1}^{l} \beta_{i1} g(\omega_i x_j + b_i) \\ \sum_{i=1}^{l} \beta_{i2} g(\omega_i x_j + b_i) \\ \vdots \\ \sum_{i=1}^{l} \beta_{im} g(\omega_i x_j + b_i) \end{bmatrix} \tag{5.20}$$

To obtain β_{jk}, Formulas (5.19) and (5.20) can be simplified to Formula (5.21), which H is the hidden layer output matrix of extreme learning machine, and the specific form is as shown in Formula (5.22). In this way, β_{jk} can be obtained by Formula (5.23), where H^+ is Moore–Penrose generalized inverse of output matrix of hidden layer.

$$H\beta = T \tag{5.21}$$

$$H = \begin{bmatrix} g(\omega_1 x_1 + b_1) & g(\omega_2 x_1 + b_2) & \cdots & g(\omega_l x_1 + b_l) \\ g(\omega_1 x_2 + b_1) & g(\omega_2 x_2 + b_2) & \cdots & g(\omega_l x_2 + b_l) \\ \vdots & \vdots & \ddots & \vdots \\ g(\omega_1 x_Q + b_1) & g(\omega_2 x_Q + b_2) & \cdots & g(\omega_l x_Q + b_l) \end{bmatrix} \tag{5.22}$$

$$\beta_{l\times m} = H^+ T \tag{5.23}$$

Finally, through the established ELM model, the corresponding output can be obtained by inputting the test set.

The establishment process of the above ELM model is shown in Fig. 5.3.

5.2.1 *Release Control Strategy for Determining Releasing Time Based on ELM*

5.2.1.1 Comparison of Simple Control Strategies for Determining Releasing Time

Simple control strategies for determining the releasing time include FIFO, EDD, CONWIP, SA, WIPCTRL and WR. In this section, we simulate on the BL model of actual semiconductor production line and compare the advantages and disadvantages of these strategies and select chip quantity and average processing cycle as performance indicators to evaluate the performance of these strategies. The model runs for 90 days, with the first 30 days as the warm-up period. The simulation results are shown in Table 5.3, in which TH_CMP and CT_CMP represent the comparison between each strategy and FIFO in th and CT performance indexes.

It can be seen from Table 5.3 that the output of FIFO and EDD in static release control strategy is equal, but the average processing cycle of EDD strategy is slightly superior to FIFO, which is reduced by 0.20%. Compared with FIFO strategy, CONWIP, SA, WIPCTRL and WR strategies have improved the chip output and average processing cycle performance by 2.70%, 3.20%, 2.96% and 3.77% respectively, and the average processing cycle by 0.30%, 3.63%, 1.01% and 3.73% respectively.

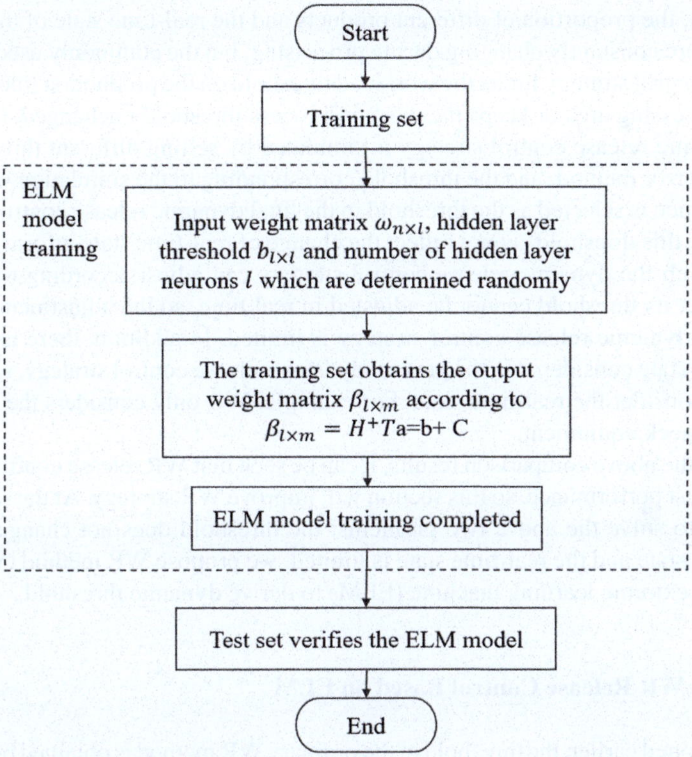

Fig. 5.3 Flow chart of extreme learning machine model construction

Table 5.3 Comparison of simple releasing strategies

Strategy	TH (lot)	TH_CMP (%)	CT (h)	CT_CMP (%)
FIFO	371	0 0.00	991	0 0.00
EDD	371	0 0.00	989	0.20
CONWIP	381	2.70	988	0.30
SA	383	3.2 0	955	3.63
WIPCTRL	382	2.96	981	1.01
WR	385	3.77	954	3.73

To sum up, the performance of dynamic release control strategy is much better than that of static release control strategy, because the dynamic release control strategy considers the real-time state of production line and can make releasing adjustment according to the real-time state, and WR strategy has the best performance among all the dynamic release control strategies.

In fact, the proportion of different products and the real-time state of the production line are constantly changing during processing, but the commonly used dynamic release control strategy limits the workload threshold on the production line or bottleneck processing area or keeps the work-in-process threshold unchanged. Generally, the dynamic release control strategy is simulated by setting different thresholds by trial and error method, and the threshold corresponding to the simulation with better performance is selected as the threshold in the final dynamic release control strategy. However, this threshold cannot reflect the change of real-time state information, that is, although the dynamic release control strategy can adjust according to the real-time state, its threshold cannot be adjusted in real time, so the adjustment made by ordinary dynamic release control strategy is limited. In addition, there is only one real-time state considered in the general dynamic release control strategy, which fails to fully consider the real-time state. For example, WR only considers the workload on bottleneck equipment.

From the above comparison results, it can be seen that WR release control strategy has the best performance, so this section will improve WR strategy. At the same time, in order to solve the above two problems: the threshold does not change with the real-time state and the real-time state is limited, we propose WR method (WRELM) based on extreme learning machine (ELM) to derive dynamic threshold.

5.2.1.2 WR Release Control Based on ELM

As mentioned earlier, the threshold in the ordinary WR method is obtained by trial and error. However, the proportion of different products invested in the actual production line is constantly changing, which will affect the constant change of real-time status. The threshold value in the ordinary WR method can not be reflected, so it is necessary to propose a WR method with real-time status threshold.

To set the dynamic threshold, it is necessary to establish a learning mechanism considering real-time state information. This learning mechanism takes multiple real-time states as input and dynamic threshold as output, which not only realizes the dynamic adjustment of threshold in WR with real-time states, but also considers multiple real-time state information. The main process of establishing learning mechanism is as shown in Fig. 5.4, specifically as follows:

Step 1: Sample collection

Firstly, WR methods with different fixed thresholds are selected for simulation. The purpose is to cover the optimal WR threshold in different real-time states as much as possible. That is to say, the threshold value in the finally generated model WRELM will dynamically change according to different real-time states, which can promote the production line performance to achieve the best.

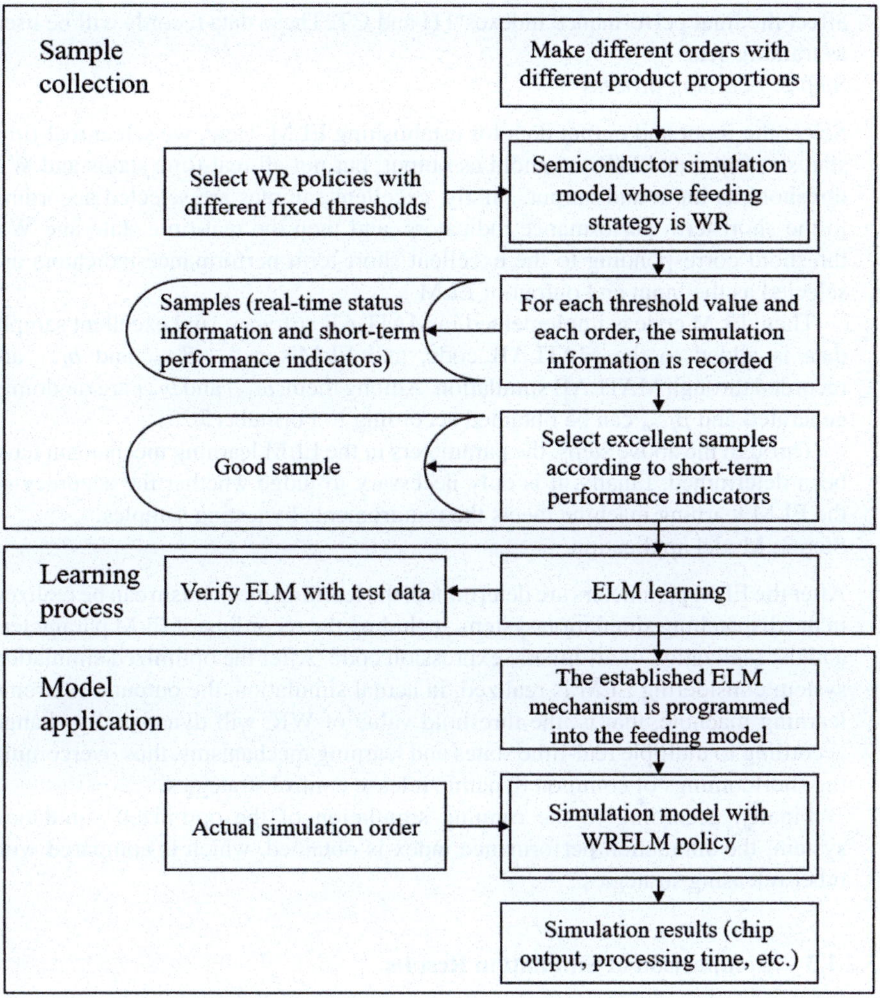

Fig. 5.4 Establishing WRELM process

Secondly, different orders with different product proportions are selected for each threshold. Putting different orders with different product proportions into the production line will make the real-time status as diversified as possible, and the learning mechanism in the final model can be applied to the ever-changing real-time status to achieve the final performance index optimization.

Finally, record real-time status information and short-term performance indicators. The real-time status information includes the number of different kinds of work-in-process and the number of work-in-process at different processing stages. Short-term performance indicators include processing steps per day (MOV per day) and utilization rate of bottleneck equipment. Short-term performance will

affect the final performance indexes TH and CT. These data records will be used as training sets.
Step 2: Learning process

Select the input and output data for establishing ELM. Here, we select real-time status as input and WR threshold as output, but not all real-time status and WR threshold as input and output. Firstly, excellent samples are selected according to the short-term performance indicators, and then the real-time state and WR threshold corresponding to the excellent short-term performance indicators are selected as the input and output of ELM.

Then, ELM code is implemented in MATLAB software, and excellent sample data is added to the MATLAB code, and ELM's $\omega_{n\times l}$, $\beta_{l\times m}$ and $b_{l\times 1}$ are recorded through MATLAB simulation. Among them, $\omega_{n\times l}$ and $b_{l\times 1}$ are randomly generated and $\beta_{l\times m}$ can be obtained according to Formula (5.23).

Through the above steps, the parameters in the ELM learning mechanism have been determined. Finally, it is only necessary to judge whether the accuracy of the ELM learning machine meets the requirements by testing samples.
Step 3: Model application

After the ELM parameters are determined, the learning mechanism can be realized in the dispatching simulation system, including the recording of ELM parameters and the realization of arithmetic expression code. After the optimized simulation system considering ELM is realized, in actual simulation, the output of extreme learning machine, that is, the threshold value of WR, will dynamically change according to multiple real-time states and learning mechanisms, thus overcoming the shortcomings of common dynamic release control strategies.

Finally, according to the running simulation of the optimized simulation system, the simulation performance index is obtained, which is compared with other releasing strategies.

5.2.1.3 Comparison of Simulation Results

The simulation here is carried out on BL model. Five different kinds of workpieces are selected in the experiment, and each type of workpiece has different processing time on the bottleneck equipment, which can ensure that orders with different proportions of workpieces can generate different real-time states on the bottleneck equipment, as shown in Table 5.4. According to the pre-simulation, the load value on the bottleneck equipment is between 0 and 900 points, and the performance result is good, so the value of 10 with an interval of 100 from 0 to 900 points is selected as different fixed thresholds in WR method. Among them, when the threshold value is 0, the order will not feed according to the workload on the bottleneck equipment, but only according to the pre-arranged order, and the releasing rule at this time is FIFO. At the same time, under each threshold, 22 orders with different proportions of workpieces were selected for 22 simulations, with one simulation for each order and 220 simulations in total.

Table 5.4 Processing times of different workpieces on bottleneck equipment

Workpiece name	Processing times of bottleneck equipment (times)
UF100300	Five
V16N50	Nine
1117F6	12
8563	15
YTD0325	22

Here, each simulation is carried out for 90 days, of which the first 30 days are the preheating period, in order to make each processing area on the production line reach a balanced state. From the 31st day, real-time status information and short-term performance index information were recorded, and 13,200 (= 60 * 220) groups of sample data were collected. Among these samples, the average daily processing steps are 3400 steps, and the utilization rate is 60%. 560 groups of samples with high daily processing steps and high utilization rate are selected as training sets, and some excellent samples with a threshold of 200 points are shown in Table 5.5.

The number of work-in-process of different workpieces and the number of work-in-process of different processing stages are taken as the input of ELM, and the corresponding threshold values are selected as the output. Then the ELM mechanism is established by MATLAB simulation, and the accuracy of the established ELM mechanism is verified by the test set test. At the same time, matrix $\omega_{n\times l}$, $b_{l\times 1}$ and $\beta_{l\times m}$ will be recorded in BL simulation system after ELM learning machine is established by MATLAB simulation.

After setting up the extreme learning machine, the advantages and disadvantages of WRELM and other simple releasing rules are compared by simulation. Because

Table 5.5 Examples of training sets

Number of processing steps per day (steps)	Utilization	Proportion of WIP quantity of different workpieces	Proportion of WIP quantity in different processing stages	Threshold value (min)
37,500	0.7025	30:33:34:62:29	21:27:26	200
39,050	0.7229	35:74:45:49:49	27:23:20	200
38,825	0.7075	34:72:41:48:48	29:21:25	200
3760	0.713 8	45:46:46:132:44	19:25:24	200
39,775	0.7135	45:46:46:131:44	86:24:20	200
35,700	0.7010	34:44:73:46:48	25:22:25	200
36,175	0.7039	55:58:74:86:59	30:22:21	200
36,100	0.7050	48:56:55:82:86	18:27:23	200
36,225	0.7055	34:37:33:46:49	30:24:22	200
36,125	0.7343	34:39:33:49:52	28:25:21	200

it is known that WR releasing strategy in simple releasing rules has the best performance, it is only compared with WR releasing strategy. Similarly, it was simulated on BL model for 90 days, and the first 30 days were used as preheating period. Two simulation results representing general performance and optimal performance are selected as comparison among several WR strategies with different thresholds. Order i($1 \leq i \leq 5$) means five orders with different workpiece proportions. The simulation results are shown in Table5.6 and the comparison of simulation results is shown in Table 5.7. To express the comparison of simulation, the simulation results are displayed in histogram, as shown in Figs. 5.5 and 5.6. WR1 and WR2 respectively represent the best simulation results and average simulation results, TH_IMP_WR1 and CT_IMP_WR1 respectively represent THe improvement percentage of WRELM compared with WR1 results th and CT, and TH_IMP_WR2 and CT_IMP_WR2 respectively represent the improvement percentage of WRELM compared with WR2 results th and CT.

The following conclusions can be drawn from Tables 5.6 and 5.7 and Figs. 5.5, 5.6:

(1) For order1 and order2, Fig. 5.5 shows that TH of WRELM is equal to TH of WR1, and Fig. 5.6 shows that CT of WRELM is slightly lower than CT of WR1, and WRELM performance is improved in CT performance index. In order3, order4 and order5, TH and CT of WRELM are better than TH and CT of WR1, but the advantages are not obvious, because WR1 has the best threshold value in WR strategy, and its performance is excellent.

Table 5.6 Comparison of simulation results between WR and WRELM

Order	WR1		WR2		WRELM	
	TH (lot)	CT (h)	TH (lot)	CT (h)	TH (lot)	CT (h)
Order1	371	338	357	348	371	325
Order2	355	348	336	382	356	324
Order3	367	319	362	345	373	300
Order4	397	320	386	500	398	317
Order5	356	324	345	500	359	321

Table 5.7 Comparison of WR and WRELM simulation results

	TH_IMP_WR1 (%)	TH_IMP_WR2 (%)	CT_IMP_WR1 (%)	CT_IMP_WR2 (%)
Order1	0 0.00	3.93	3.84	6.61
Order2	0 0.00	5.95	6.90	15.18
Order3	1.63	3.04	5.96	13.04
Order4	0.23	3.11	0.94	20.75
Order5	0.84	4.05	0.93	19.75

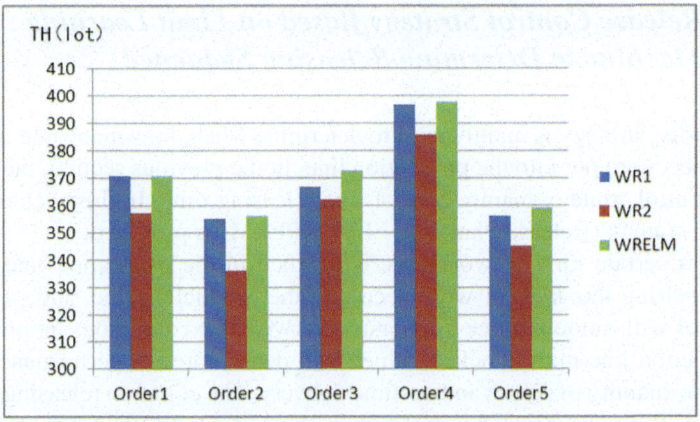

Fig. 5.5 Comparison of the influence of different releasing methods on TH performance

Fig. 5.6 Comparison of the influence of different releasing methods on CT performance

(2) Compared with WR2, the simulation performance of WRELM strategy is obviously improved. In several simulations, WRELM has a maximum improvement range of 5.94% and a minimum improvement range of 3.04% compared with WR2. On CT, the maximum value was 20.75%, and the minimum value was 6.61%.

(3) On the whole, the performance of WRELM is similar to that of WR1 with the best performance, but compared with WR2 with general performance, it can be improved by 4.03% on TH and 15.40% on CT, respectively.

5.2.2 Release Control Strategy Based on Limit Learning Machine to Determine Releasing Sequence

The releasing strategy is mainly used to determine when, how many and what kinds of workpieces are put into the production line. In the previous section, the WRELM release control strategy mainly controls the releasing time. In this section, another releasing strategy (Release Plan with ELM, RPELM) is proposed.

When a certain kind of workpiece is blocked in the production line, we need to delay putting this kind of workpiece into the production line, and put in those workpieces with smooth processing first, otherwise, the continuous input will make the production line more blocked, which will deteriorate the performance indexes such as the quantity of pieces and on-time delivery. The common releasing strategies considering the releasing sequence mainly include FIFO and EDD, but only the order information is considered, ignoring the real-time status information on the production line. RPELM considers the order information and the real-time status information of the production line at the same time to decide the releasing sequence.

5.2.2.1 Analysis of Order Information Affecting Releasing

For different workpieces, the inherent attributes of workpieces mainly include the number of processing steps, net processing time, the average processing cycle given by the order and whether they are urgent workpieces. The number of processing steps of a workpiece indicates the length of the workpiece flow. The more processing steps, the more times it is dispatched, which will increase the waiting time of the workpiece in the processing area and affect the average processing cycle of the workpiece. The net processing time means the time required for the workpiece to finish processing without blockage. The smaller the net processing time means that the residence time of the workpiece on the production line is less, and the average processing cycle is less when the production line is smooth. The average processing cycle indicates the average time interval between the start of processing and the completion of processing for a type of workpiece. The smaller the average processing cycle, the faster its fluidity, the higher the utilization rate of equipment and the higher the productivity of the production line. If a workpiece is an urgent workpiece, it should be put into the production line first to optimize the on-time delivery rate (HLODR) and the average processing cycle (CT) of the urgent work-piece.

To sum up, there are four factors that affect the order information of releasing: net processing time, average processing cycle of the order, processing steps and whether it is urgent or not.

5.2.2.2 Dispatching Rules with Emergency Workpieces

Because the urgent work piece is considered, it is necessary to distinguish the urgent work piece from the ordinary work piece in the simulation model. Emergency workpieces are those that need to be put into the production line first due to tight delivery time, so the delivery time of emergency workpieces is set to be smaller than that of ordinary workpieces. Here, the delivery time of ordinary workpieces is defined as the input time plus the average processing cycle of workpieces, and the emergency delivery time of workpieces is the product of the input time plus the average processing cycle of workpieces and a coefficient which is a random number between 0.8 and 1. The delivery time of ordinary workpieces and urgent workpieces is shown in Formulas (5.24) and (5.25).

$$Due\ Date(i) = Release\ Time(i) + CT(i) \tag{5.24}$$

$$HotLotDueDate(i) = Release\ Time(i) + CT(i) * random(0.8, 1) \tag{5.25}$$

where, $DueDate(i)$ is the delivery time of the workpiece i, $HotLotDueDate(i)$ is the delivery time of the emergency workpiece i, $Release\ Time(i)$ is the releasing time of the workpiece i, $CT(i)$ is the given processing time of the workpiece i, and $random(0.8, 1)$ is a random number ranging from 0.8 to 1.

We do not consider the complicated dispatching rules when making the release control strategy, but simply adopt FIFO, which means that the workpieces entering the processing area first are processed first, without considering the urgency of the workpieces. However, urgent workpieces have been added to the production line, so the dispatching rules must consider the urgent workpieces and make some changes. Here, we improve the FIFO dispatching rule, so that the dispatching rule first observes whether there are urgent workpieces in the processing area, and if there are urgent workpieces, according to the FIFO rule, select the earliest workpiece in the processing area for processing; if there is no urgent workpiece, select the workpiece for processing according to FIFO rules.

In the same situation, there are batch processing rules. Generally, in the processing area where batch processing equipment is located, a class of workpieces will be selected according to FIFO for batch processing. However, in the case of emergency workpieces, the batch processing rules will first bring a class of emergency workpieces into the batch range according to FIFO. If the number of emergency workpieces is less than the number of batch workpieces, the same kind of common workpieces will be selected according to FIFO to be combined with emergency workpieces. If there is no urgent work piece, simply select one type of work piece according to FIFO for batch processing.

To compare the improved dispatching rules with those before the improvement, we simulated on MIMAC simulation model when the proportion of urgent parts was 10% and the number of products was 2500. The release control strategies in the simulation are FIFO, EDD and RPELM, which will be described in detail below. The

Table 5.8 Simulation comparison before and after improving dispatching rules

	Simulation results before improving dispatching rules			Simulation results after improving dispatching rules		
	FIFO	EDD	RPELM	FIFO	EDD	RPELM
TH (lot)	582	587	586	582	581	582
CT (h)	351	348	347	356	354	351
ODR (%)	46.76	47.43	50.67	48.77	48.76	52.41
HLODR (%)	37.5 0	57.5 0	52.76	90.70	100	100

Table 5.9 Simulation performance change ratio before and after dispatching rules change

	FIFO (%)	EDD (%)	RPELM (%)
TH(lot)	0 0.00	– 1.02	– 0.68
CT (h)	– 1.43	– 1.72	– 1.15
ODR	2.01	1.33	1.74
HLODR	53.20	42.50	48.50

main purpose of the simulation is to observe the simulation performance changes before and after the change of dispatching rules. The simulation lasts for 90 days, of which the first 30 days are the preheating period. The simulation results are shown in Table 5.8, and the performance change ratio of the simulation results is shown in Table 5.9.

From Table 5.9, we can see that TH performance index is slightly worse but not obvious before and after the change of dispatching rules. The performance index of CT is obviously deteriorated, which is reduced by – 1.43%, – 1.72% and – 1.15% respectively when the releasing rules are FIFO, EDD and RPELM. However, the performance of ODR and HLODR has been improved, especially the performance of HLODR has been greatly improved. Under the releasing rules of FIFO, EDD and RPELM, ODR was improved by 2.01%, 1.33% and 1.74% respectively. While HLODR was improved by 53.2 0%, 42.5 0% and 48.5 0% respectively. The improved dispatching rules can play an obvious role in improving the delivery rate of urgent workpieces, and the advantages far outweigh the disadvantages in terms of the overall performance index. Therefore, in MIMAC simulation, when urgent workpieces are involved, the improved dispatching rules will be adopted.

5.2.2.3 Determination of Releasing Priority by Multiple Linear Regression Equation

The order factors that affect the releasing include net processing time, average processing time of the order, processing steps and whether it is urgent or not. To determine the relationship between these four factors and releasing sequence, multiple linear regression equation can be used to determine the releasing priority of workpieces.

Linear regression is an expression that has been deeply studied and widely used in practical applications, because the model that linearly depends on its unknown parameters is easier to fit than the model that nonlinearly depends on its position parameters, and the statistical characteristics produced are easier to determine.

The general mathematical expression of linear regression is:

$$Y = \alpha_1 x_1 + \alpha_2 x_2 + \ldots + \alpha_m x_m \tag{5.26}$$

In Formula (5.26), $\alpha_1, \alpha_2, \ldots, \alpha_m$ are partial regression coefficients.

For ordinary workpieces, Formula (5.26) can be used to express the relationship between the releasing priority of workpieces and the order information table, but for urgent workpieces, an additional parameter is needed to further improve the releasing priority of urgent workpieces, as shown in Formula (5.27).

$$P_i = \text{a}*\frac{CT_i}{\max(CT_i)} + b * \frac{T_i}{\max(T_i)} + c * \frac{Steps_i}{\max(Steps_i)} + IsHotLot(i) \tag{5.27}$$

where P_i, CT_i, T_i and $Steps_i$ respectively represent the workpiece priority, the average processing cycle given by the order, the net processing time given by the order and the processing steps of the workpiece Lot_i. The value of $IsHotLot(i)$ in Formula (5.27) is determined according to whether the workpiece Lot_i is an emergency workpiece. If the workpiece is an emergency workpiece, the value is 1; otherwise, the value is 0. a, b and c respectively represent the weights of CT_i, T_i and $Steps_i$ of the workpiece Lot_i, which are derived by ELM learning mechanism with real-time status information as input, so RPELM considers both order information and real-time status information

5.2.2.4 Determine the Releasing Sequence Based on ELM

In the previous section, we have obtained the multiple linear regression equation of order information and workpiece releasing priority and proposed that in order to deduce the weight of workpiece information from real-time status, we need to establish an extreme learning machine. To establish ELM learning machine, we need to obtain enough samples and select excellent samples as training sets. Training set is the input and output data of ELM which is not established.

The construction process of simulation model for determining releasing sequence based on ELM and multiple linear regression equation is shown in Fig. 5.7, and the main steps are as follows.

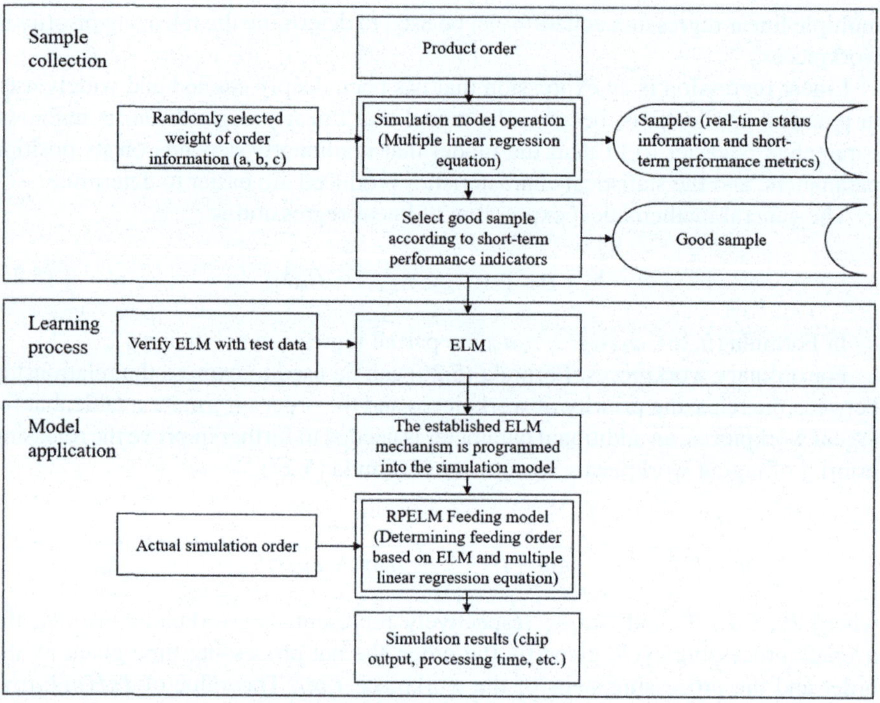

Fig. 5.7 Flow chart of constructing RPELM releasing model

Step 1: Sample collection

First, a, *b*, and *c* are selected randomly as weight coefficients. The randomly generated a, *b* and *c* are applied to the multivariate linear regression equation to obtain the workpiece releasing priority, and then the samples are generated by simulation according to the workpiece releasing priority.

Then, you need to record the sample. Sample data refers to real-time status information and short-term performance indicators. Real-time status information includes the number of front, middle and back stages of different kinds of workpieces in the production and processing stage. Short-term performance indicators include utilization rate of bottleneck equipment, daily processing steps (MovPerDay), daily sheet output (THPerDay) and monthly on-time delivery rate (ODRPerDay). These short-term performance indicators can reflect the long-term performance indicators, that is, the final performance indicators, including TH, CT and ODR, HLODR and so on.

Step 2: Learning process

First, select ELM's training set. The real-time status corresponding to excellent short-term performance indicators and the statistical results of order information specific gravity values are selected as the input and output of ELM respectively. Then, programming on MATLAB, simulating running and establishing the extreme learning mechanism.

Secondly, the set of test samples is used to test the established learning mechanism to check whether the accuracy of ELM learning mechanism meets the requirements.

Step 3: Model application

Implement learning mechanism in scheduling simulation system. Includes the record of ELM parameters and the implementation of arithmetic expression code. After the above steps are completed, the priority of the order artifacts in the optimized simulation system will be changed in real time according to the order information, real-time status information and ELM mechanism.

Finally, according to the simulation of the optimized simulation system, the simulation performance index is obtained and compared with other releasing strategies.

5.2.2.5 Comparison of Simulation Results

To compare the advantages and disadvantages of the releasing strategies of RPELM, FIFO and EDD, different strategies of FIFO, EDD and RPELM were simulated on MIMAC model, and the simulation results were compared. Each simulation lasts for 90 days, in which the first 30 days are the warm-up period, and the dispatching rule is the improved FIFO.

There are 9 kinds of workpieces in MIMAC model, and the number of them in the front, middle and back stages of machining is selected as real-time status, with 27 real-time status information. At THe same time, th, CT, ODR and HLODR are selected as performance evaluation indexes.

To make the results more convincing, simulation was carried out under the condition that the number of products in production line was fixed at 2500 pieces, 3500 pieces, 4500 pieces and 5500 pieces respectively. The simulation results are shown in Table 5.10, and the comparison of simulation results is shown in Table 5.11. For the convenience of observing the results, the simulation results are displayed as histogram, as shown in Figs. 5.8, 5.9, 5.10 and 5.11.

The following conclusions can be drawn from Tables 5.10 and 5.11 and Figs. 5.8, 5.9, 5.10 and 5.11:

(1) From Fig. 5.8, the TH of various strategies has hardly changed under the same WIP quantity restriction, and the biggest change range between strategies is that RPELM is improved by 1% compared with EDD at 5500 pieces, which is due to the fact that the WIP quantity limits the processing capacity of the production line, thus limiting the output of the production line. At the same

Table 5.10 Simulation results of FIFO, EDD and RPELM releasing strategies under different WIP quantities

Workload (unit)	FIFO				EDD				RPELM			
	TH (lot)	CT (h)	ODR (%)	HLODR (%)	TH (lot)	CT (h)	ODR (%)	HLODR (%)	TH (lot)	CT (h)	ODR (%)	HLODR (%)
2500	582	356	48.77	90.70	581	354	48.76	100	582	356	52.41	100
3500	674	408	49.53	750.00	671	409	48.91	71.43	671	408	52.85	78.57
4500	646	635	48.79	76.67	644	654	48.74	76.67	646	634	52.51	78.58
5500	704	831	47.62	54.16	698	842	48.01	54.16	705	829	49.79	54.16

Table 5.11 Comparison of simulation results of FIFO, EDD and RPELM releasing strategies under different WIP quantities

Workload (unit)	Improvement of performance of RPELM relative to FIFO				Improvement of performance of RPELM relative to EDD			
	TH (lot) (%)	CT (h) (%)	ODR (%)	HLODR (%)	TH (lot) (%)	CT (hour) (%)	ODR (%)	HLODR (%)
2500	0.00	0.00	3.64	9.30	0.17	– 0.57	3.65	0.00
3500	– 0.45	0.00	3.32	3.57	0.00	0.24	3.94	7.14
4500	0.00	0.16	3.72	1.91	0.31	3.06	3.77	1.91
5500	0.14	0.24	2.17	0.00	1.00	1.54	1.78	0.00

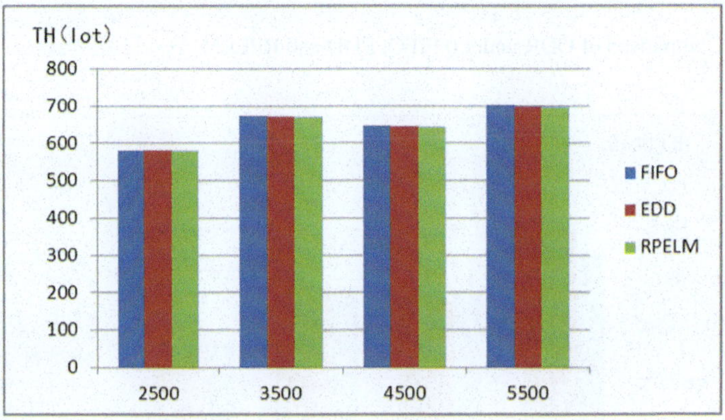

Fig. 5.8 Comparison of TH under FIFO, EDD and RPELM

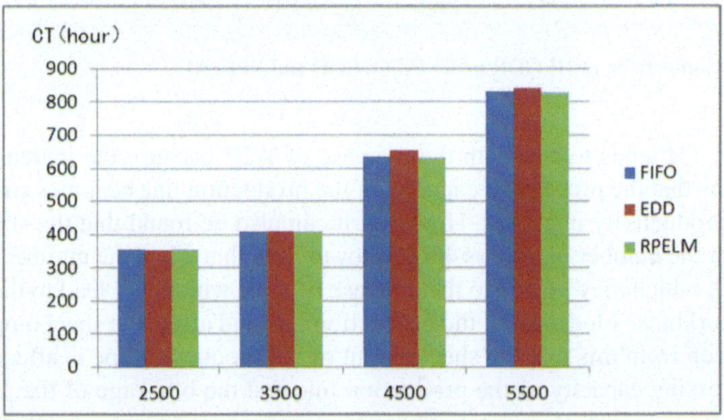

Fig. 5.9 Comparison of CT under FIFO, EDD and RPELM

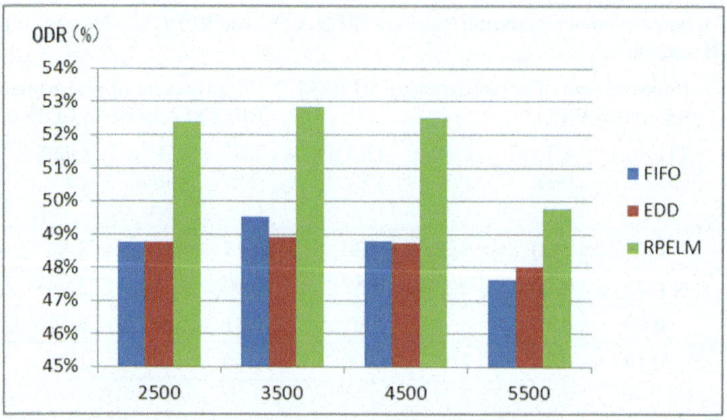

Fig. 5.10 Comparison of ODR under 0 FIFO, EDD and RPELM

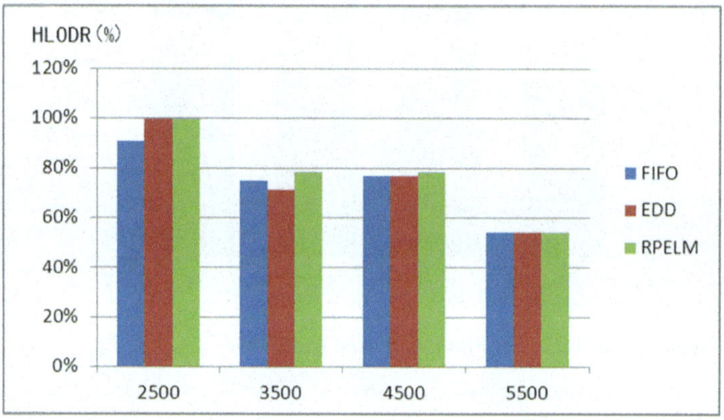

Fig. 5.11 Comparison of HLODR under FIFO, EDD and RPELM

time, TH tends to rise with the increase of WIP, because the increase of WIP means that the processing capacity of the production line becomes stronger and the productivity increases. However, it can also be found that the sheet output when the number of WIP is 4500 is lower than that when the number of WIP is 3500, which may be due to the increase of WIP, which will lead to the increase of workpiece blockage on the production line and affect the sheet output. It can be seen from this that the sheet output of the production line is affected by the processing capacity of the production line and the blockage of the production line.

(2) It can be seen from Fig. 5.9 that the CT of RPELM is improved compared with that of FIFO and EDD, but it is not obvious. At the same time, it can be found that with the increase of the number of fixed-in-process products, the CT performance under various releasing strategies decreases consistently.

This is because the increase of the number of fixed-in-process products in the production line will aggravate the blockage of workpiece processing in the production line, and the waiting time of workpieces in the processing area will be prolonged as a whole, resulting in the decline of CT performance index.

(3) Fig. 5.10 shows that ODR can be improved a lot by using RPELM release control strategy. With 2500 pieces, 3500 pieces, 4500 pieces and 5500 pieces of fixed products, the RPELM can be increased by 3.64%, 3.32%, 3.72% and 2.17% compared with FIFO and 3.65%, 3.94%, 3.77% and 1.78% compared with EDD, respectively. This shows that RPELM can make real-time policy adjustments to improve ODR by considering real-time status information. At the same time, however, we can also find that ODR performance decreases with the increase of the number of fixed products, which is also caused by the increased blockage of the production line.

(4) Fig. 5.11 shows that when the load of the production line is light, the delivery time of emergency workpieces can be obviously improved by RPELM. Compared with FIFO, RPELM improves HLODR performance by 9.3 0%, 3.57% and 1.91%, respectively, when the number of fixed WIP is 2500 pieces, 3500 pieces and 4500 pieces. Compared with EDD, RPELM increased by 7.14% and 1.91%, respectively, when the number of fixed products was 3500 and 4500. At the same time, we can also find that the performance improvement of HLODR will weaken with the increase of the number of WIP in the production line. The reason is also that the blockage of the production line intensifies, which makes the delivery time of urgent workpieces decrease.

Overall, RPELM can achieve better final performance index than FIFO and EDD. Especially when the production line is light load, RPELM can effectively improve the on-time delivery rate of workpieces and urgent workpieces. Obviously, this is due to the real-time adjustment made by RPELM considering the real-time state of the production line.

5.3 Optimization of Release Control Based on Attribute Selection

In the RPELM strategy, the number of different kinds of workpieces in different processing stages is directly selected as the real-time state, because these are important real-time states for changing the releasing sequence according to experience. But in fact, there are many attributes that will affect the releasing sequence. Because many attributes are highly correlated, and some attributes have no influence on performance indicators, we need to select representatives from a highly correlated set of attributes and eliminate those attributes that have no great influence on performance indicators from the set of attributes.

5.3.1 Attribute Set Related to Releasing

Due to the complex characteristics of semiconductor manufacturing system, there are many real-time states on the production line. There are many attributes that affect the releasing strategy, so it is necessary to mine the real-time status attributes that have great influence on the release control strategy. Firstly, it is necessary to select the attribute set on the production line. In addition to the number of different workpieces before, during and after the production line, it also selects the number of WIP, processing steps, processing time, processing steps on bottleneck equipment and the number of workpieces waiting to be processed on bottleneck equipment as real-time status information. The attribute sets embodied in the MIMAC simulation model are shown in Tables 5.12, 5.13, 5.14, 5.15, 5.16, 5.17, 5.18, 5.19 and 5.20.

5.3.2 Attribute Selection

Attribute selection is a process of selecting some of the most effective features from all features to reduce the dimension of feature space, which mainly includes four basic steps: generation of candidate feature subsets (search strategy), evaluation criteria, stopping criteria and verification methods. The basic methods of attribute selection are as follows:

(1) Mean square error evaluation method

Find the difference between the measured values of each comparison column (non-standard column) and the standard column, and then sum the squares of the differences, and then average them. The mean square error calculation formula is:

$$R_k = \frac{1}{n}\sum_{i=1}^{n}(x_{ki} - x_{0i})^2 \tag{5.28}$$

where, x_0 is standard data and n is the number of valid data. The smaller the value of R_k, the smaller the difference between the non-standard data and the standard data.

(2) Spectrum analysis method

Firstly, the standard data and non-standard data are Fourier transformed, and then the mean value of the square of the difference between the amplitude values of each non-standard data and standard data is calculated.

$$R_k = \frac{1}{n}\sum_{i=1}^{n}(fft(x_{ki}) - fft(x_{0i}))^2 \tag{5.29}$$

Table 5.12 Production line attribute set

Serial number	Attribute name	Attribute meaning
1	WIPPerDay	WIP of production line
2	MovPerDay	Production line Mov
3	ThPerDay	Output of production line
4	ProTimrPerDay	Processing time of production line
5	PreWIP1	The first 1/3 of the first product in-process quantity
6	PreWIP2	The first 1/3 of the second product in-process quantity
7	PreWIP3	The first 1/3 of the third product in-process quantity
8	PreWIP4	The first 1/3 of the fourth product in-process quantity
9	PreWIP5	The first 1/3 of the fifth product in-process quantity
10	PreWIP6	The first 1/3 of the sixth product in-process quantity
11	PreWIP7	The first 1/3 of the seventh product in-process quantity
12	PreWIP8	The first 1/3 of the eighth product in-process quantity
13	PreWIP9	Number of work-in-process of the first 1/3 ninth product
14	MidWIP1	1/3 of the first product in-process quantity
15	MidWIP2	1/3 of the second product in-process quantity
16	MidWIP3	1/3 of the third product in-process quantity
17	MidWIP4	1/3 of the fourth product in-process quantity
18	MidWIP5	1/3 of the fifth product in-process quantity
19	MidWIP6	1/3 of the sixth product in-process quantity
20	MidWIP7	1/3 of the 7th product in-process quantity
21	MidWIP8	1/3 of the eighth product in-process quantity
22	MidWIP9	1/3 of the ninth product in-process quantity
23	BehWIP1	After 1/3 of the first product in-process quantity
24	BehWIP2	After 1/3 of the second product, the number of WIP
25	BehWIP3	After 1/3 of the third product in-process quantity
26	BehWIP4	After 1/3 of the fourth product in-process quantity
27	BehWIP5	After 1/3 of the fifth product in-process quantity
28	BehWIP6	After 1/3 of the sixth product in-process quantity
29	BehWIP7	After 1/3, the number of work-in-process of the seventh product
30	BehWIP8	After 1/3, the number of work-in-process of the eighth product
31	BehWIP9	After 1/3, the number of work-in-process of the ninth product

Table 5.13 Buffer11021_ASM_A1_A3_G1 processing zone attribute set

Serial number	Attribute name	Attribute meaning
32	Mov	Buffer11021_ASM_A1_A3_G1 Mov in processing area
33	Queue	Buffer11021_ASM_A1_A3_G1 queue leader of processing area
34	Utilization	Buffer11021_ASM_A1_A3_G1 Processing Zone Utilization Rate

Table 5.14 Buffer1024_ASM_A4_G3_G4 processing zone attribute set

Serial number	Attribute name	Attribute meaning
35	Mov	Buffer1024_ASM_A4_G3_G4 Mov in machining area
36	Queue	Buffer1024_ASM_A4_G3_G4 Queue Leader in Processing Zone
37	Utilization	Buffer1024_ASM_A4_G3_G4 Processing Area Utilization Rate

Table 5.15 Buffer11026_ASM_B2 processing zone attribute set

Serial number	Attribute name	Attribute meaning
38	Mov	Buffer11026_ASM_B2 processing zone Mov
39	Queue	Buffer11026_ASM_B2 queue leader of processing zone
40	Utilization	Buffer11026_ASM_B2 utilization rate of processing area

Table 5.16 Buffer11027_ASM_B3_B4_D4 processing area attribute set

Serial number	Attribute name	Attribute meaning
41	Mov	Buffer11027_ASM_B3_B4_D4 Mov in processing area
42	Queue	Buffer11027_ASM_B3_B4_D4 queue leader of processing area
43	Utilization	Buffer11027_ASM_B3_B4_D4 processing area utilization rate

Table 5.17 Buffer11029_ASM_C1_D1 processing zone attribute set

Serial number	Attribute name	Attribute meaning
44	Mov	Buffer11029_ASM_C1_D1 Mov in processing area
45	Queue	Buffer11029_ASM_C1_D1 queue leader of processing area
46	Utilization	Buffer11029_ASM_C1_D1 Processing Zone Utilization Rate

Table 5.18 Buffer11030_ASM_C2_H1 processing zone attribute set

Serial number	Attribute name	Attribute meaning
47	Mov	Buffer11030_ASM_C2_H1 C2 _ H1MOV in processing area
48	Queue	Buffer11030_ASM_C2_H1 Queue Leader of Processing Zone
49	Utilization	Buffer11030_ASM_C2_H1 Processing Area Utilization Rate

Table 5.19 Buffer17221_K_SMU236 processing zone attribute set

Serial number	Attribute name	Attribute meaning
50	Mov	Buffer17221_K_SMU236 processing area Mov
51	Queue	Buffer17221_K_SMU236 queue leader of work area
52	Utilization	Buffer17221_K_SMU236 utilization rate of processing area

Table 5.20 Buffer17421_HOTIN processing zone attribute set

Serial number	Attribute name	Attribute meaning
53	Mov	Buffer17421_HOTIN processing area Mov
54	Queue	Buffer17421 _ queue leader of hot in processing zone
55	Utilization	Buffer17421_HOTIN processing area utilization rate

where, x_0 is standard data and n is the number of valid data. The smaller the value of R_k, the smaller the difference between the non-standard data and the standard data.

(3) Evaluation of correlation coefficient

Calculation formula of correlation coefficient:

$$\rho_{XY} = Cov(X, Y)/\left(\sqrt{D(X)}\sqrt{D(Y)}\right) \tag{5.30}$$

Among them,

$$Cov(X, Y) = E((X - E(X))(Y - E(Y))) \tag{5.31}$$

$$\mathrm{D(X)} = \mathrm{E}\left((\mathrm{X} - \mathrm{E(X)})^2\right) = \mathrm{E}\left(\mathrm{X}^2\right) - (\mathrm{E(X)})^2 \tag{5.32}$$

Covariance and variance of x and y respectively. Correlation coefficient method can express the correlation between two columns of data, and the closer its value is to 1, the closer the data are.

(4) Evaluation method of goodness of fit

According to the evaluation standard of least square data fitting, this paper tries to use its goodness-of-fit evaluation parameter R^2 to evaluate.

The formula for calculating the goodness of fit R^2 is:

$$R^2 = 1 - SSE/SST \tag{5.33}$$

Among them,

$$SSE = \sum_{i=1}^{n} (X_{ki} - X_{0i})^2 \tag{5.34}$$

$$\mathrm{SST} = \sum_{i=1}^{n} X_{0i}^2 - \frac{1}{n}\left(\sum_{i=1}^{n} X_{ki}\right)^2 \tag{5.35}$$

The larger R^2 is, the better the fitting effect is.

Use non-standard column data to fit standard data and evaluate according to the goodness of fit evaluation standard. The closer the value is to 1, the closer the column is to the standard value.

In addition to the above basic attribute selection algorithms, there are some improved attribute selection algorithms, such as attribute selection algorithm based on support vector machine pre-classification, correlation attribute selection algorithm based on maximal connected subgraph, attribute selection algorithm based on fractal dimension and ant colony algorithm, and attribute selection algorithm based on kernel function parameter optimization.

5.3.3 Simulation Based on Attribute Selection

5.3.3.1 Simulation Verification Based on MIMAC

In order to ensure the excellent performance index, it is necessary to select the production line attributes closely related to the performance index as the real-time status. Here, to ensure the advantages of the RPELM method in each performance index, firstly, the attribute set with the greatest correlation with the releasing performance index TH is selected as the real-time state set. For example, 25 final real-time states are selected on MIMAC, and the results selected by correlation coefficient method, mean square error evaluation method, spectrum analysis method and goodness-of-fit evaluation method are shown in Table 5.21.

The serial numbers in Table 5.21 refer to the attributes corresponding to the serial numbers in Tables 5.12, 5.13, 5.14, 5.15, 5.16, 5.17, 5.18, 5.19 and 5.20 respectively.

Table 5.21 Attribute set after attribute selection

Correlation coefficient method	3 4 5 6 7 12 13 23 24 25 26 27 28 29 32 33 35 41 43 46 49 50 51 54 55
Mean square error evaluation method	4 6 7 8 9 10 11 12 13 15 16 17 18 20 21 22 25 26 27 28 29 30 38 49 55
Spectrum analysis method	4 6 7 8 9 10 11 12 13 15 16 17 18 21 22 25 26 27 28 29 30 38 49 50 55
Evaluation method of goodness of fit	4 6 7 8 9 10 11 12 13 15 16 17 18 20 21 22 25 26 27 28 29 30 38 49 55

It can be found that the attribute sets selected by the mean square error evaluation method and the goodness-of-fit evaluation method are exactly the same, and most of the attributes selected by the spectrum analysis method and the attribute set selected by the mean square error are coincident. Although the attribute set selected by the correlation coefficient method is different from the attribute sets selected by the other three attribute selection algorithms, it is basically the same.

In order to compare the effectiveness of these four methods, the attributes selected by these four attribute selection methods are applied to the releasing strategy of RPELM, and the performance results are recorded by simulation. At the same time, we also take FIFO and EDD as the performance indicators after the releasing strategy simulation operation as the comparison objects. Because the attribute sets selected by the mean square error evaluation method and the goodness of fit evaluation method are the same here, only the RPELM after the attribute selection by the mean square error method is simulated. The simulation runs for 320 days, of which the first 30 days are used as the preheating period. The simulation results are shown in Table 5.22, and the comparison of simulation results is shown in Table 5.23. RPELM_FS represents rpelm releasing strategy considering attribute selection.

The following conclusions can be drawn from Tables 5.22 and 5.23:

(1) From Table 5.22, it can be seen that the performance index of RPELM_FS method after attribute selection has been improved compared with FIFO and EDD. RPELM_FS considering correlation coefficient method has obvious improvement effect on ODR and HLODR, which is 3.02% and 2.01% respectively compared with FIFO, 2.26% and 1.00% respectively compared with EDD, and TH and CT also have some improvements.

(2) Compared with FIFO and EDD, RPELM_FS considering the mean square error method and spectrum analysis method can also improve HLODR, and the improvement range of HLODR is greater than that of correlation coefficient method. Especially, compared with FIFO and EDD, RPELM after spectrum analysis can improve HLODR by 4.53% and 3.52%, and the improvement effect is obvious. However, the improvement of ODR by these two methods is not as obvious as that of correlation coefficient method.

Table 5.22 Simulation results of FIFO, EDD and RPELM after attribute selection

	FIFO	EDD	RPELM	RPELM_FS		
				Correlation coefficient method	Mean square error evaluation method	Spectrum analysis method
TH (lot)	2396	2393	2394	2398	2398	2395
CT (h)	352	351	351	350	350	351
ODR (%)	25.39	26.15	26.41	28.41	26.74	26.71
HLODR (%)	45.22	46.23	46.97	47.23	47.74	49.75

Table 5.23 Somparison of RPELM strategy with FIFO and EDD after attribute selection

	RPELM_FS								
	Correlation coefficient method			Mean square error evaluation method			Spectrum analysis method		
Polices	FIFO (%)	EDD (%)	RPELM (%)	FIFO (%)	EDD (%)	RPELM (%)	FIFO (%)	EDD (%)	RPELM(%)
TH (lot)	0.08	0.21	0.17	0.08	0.01	0.17	– 0.04	0.08	0.04
CT (h)	0.57	0.28	0.28	0.57	0.28	0.28	0.28	00.00	00.00
ODR	3.02	2.26	1.69	0.35	0.59	0.33	1.32	0.56	0.3 0
HLODR	2.01	1.00	0.26	2.52	1.51	0.77	4.53	3.52	2.78

(3) Compared with RPELM, RPELM_FS can also improve the performance index, but it is not obvious. Compared with RPELM, RPELM_FS considering correlation coefficient method has improved greatly in ODR, with an increase of 1.69%; compared with RPELM, RPELM_FS, which considers spectrum analysis method, has a great improvement on HLODR, with an increase of 2.78%.

5.3.3.2 Simulation Verification Based on BL

In order to verify the validity of RPELM after attribute selection, we also carry out simulation verification on BL model. Each simulation lasts for 300 days, including 30 days of warm-up period. At the same time, the simulation also adopts different dispatching rules, including FIFO, EDD, SPT, LPT, SPRT and ls, so that we can determine which dispatching rule RPELM_FS is suitable for.

Here, the number of different workpieces in different processing stages, daily WIP, daily processing steps, daily production time, processing steps on bottleneck equipment and queue length on bottleneck equipment are selected as attribute sample sets, with a total of 38 attributes. Every time we simulate RPELM_FS, we use correlation coefficient method to select 9 attributes as real-time status. At THe same time, th (film output), CT (average processing time), VAR (the variance of CT), HLCT (the CT of hot-lots, Average processing cycle of urgent workpiece), hlvar (the variance of the hot-lots' CT), CLCT (the CT of common-lots, Common workpiece processing time), clvar (THe variance of the common-lots' CT), ODR (delivery time), and HLODR (emergency workpiece delivery time) are taken as performance indicators, among which we focus on th, CT, ODR and HLODR. The simulation results are shown in Tables 5.24, 5.25, 5.26, 5.27, 5.28 and 5.29. In which C_FIFO and C_EDD respectively represent the comparison of simulation results of RPELM_FS with FIFO and EDD after considering different attribute selection algorithms.

The following conclusions can be drawn from Table 5.24, 5.25, 5.26, 5.27, 5.28 and 5.29:

Table 5.24 Comparison of simulation results with FIFO dispatching rule

	FIFO	EDD	RPELM_FS	C_FIFO (%)	C_EDD (%)
TH (lot)	2065	2070	2070	0.24	00.00
CT (h)	1150	1148	1148	0.17	00.00
VAR	435.62	434.08	433.42	0.51	0.15
HLCT (h)	1135	1131	1131	0.35	00.00
HLVAR	363.80	361.93	361.82	0.54	0.03
CLCT (h)	1150	1149	1149	0.09	00.00
CLVAR	443.22	441.64	440.92	0.52	0.16
ODR (%)	46.79	48.63	48.82	1.03	0.19
HLODR (%)	20.95	24.76	27.62	6.67	2.86

Table 5.25 Comparison of simulation results with EDD dispatching rule

	FIFO	EDD	RPELM_FS	C_FIFO (%)	C_EDD (%)
TH (lot)	2007	2001	1996	− 0.55	− 0.25
CT (h)	865	859	860	0.58	− 0.12
VAR	661.22	641.77	649.56	1.76	− 1.21
HLCT (h)	802	804	802	00.00	0.25
HLVAR	537.49	535.09	531.39	1.10	0.69
CLCT (h)	870	864	865	0.57	− 0.12
CLVAR	672.35	651.52	659.89	1.85	− 1.13
ODR (%)	57.54	58.91	59.31	1.77	0.40
HLODR (%)	69.39	68.70	67.34	− 2.05	− 1.36

Table 5.26 Comparison of simulation results with SPT dispatching rule

	FIFO	EDD	RPELM_FS	C_FIFO (%)	C_EDD (%)
TH (lot)	2085	2080	2084	− 0.05	0.19
CT (h)	979	973	975	0.41	− 0.21
VAR	374.15	377.71	373.0625	0.29	1.23
HLCT (h)	1025	1018	1020	0.49	− 0.20
HLVAR	459.84	448.21	452.61	1.57	− 0.98
CLCT (h)	973	967	969	0.41	− 0.21
CLVAR	364.37	369.49	363.92	0.12	1.51
ODR (%)	42.39	43.51	43.54	1.15	0.03
HLODR (%)	37.56	40.64	40.64	3.08	0.00

Table 5.27 Comparison of simulation results with LPT dispatching rule

	FIFO	EDD	RPELM_FS	C_FIFO (%)	C_EDD (%)
TH (lot)	2048	2047	2044	− 0.20	− 0.15
CT (h)	973	974	975	− 0.21	− 0.10
VAR	383.26	382.84	382.88	0.10	− 0.01
HLCT (h)	964	973	974	− 1.04	− 0.10
HLVAR	367.04	364.02	367.59	− 0.15	− 0.98
CLCT (h)	975	974	975	0.00	− 0.10
CLVAR	384.91	384.2	384.79	0.03	− 0.15
ODR (%)	48.73	49.26	48.80	0.07	− 0.46
HLODR (%)	30.90	31.43	33.71	2.81	2.28

Table 5.28 Comparison of simulation results with SPRT dispatching rule

	FIFO	EDD	RPELM_FS	C_FIFO (%)	C_EDD (%)
TH(lot)	2387	2381	2387	0.00	0.25
CT(h)	701	709	705	– 0.57	0.56
VAR	285.50	307.45	288.25	– 0.88	6.24
HLCT(h)	583	613	611	– 4.80	0.33
HLVAR	227.46	273.01	230.80	– 1.47	15.46
CLCT(h)	717	721	717	0. 00	0.55
CLVAR	291.6	301.64	294.44	– 0.97	2.39
ODR (%)	45.32	48.78	43.26	– 2.06	– 5.52
HLODR (%)	52.42	54.20	48.22	– 4.20	– 5.98

Table 5.29 Comparison of simulation results with LS dispatching rule

	FIFO	EDD	RPELM_FS	C_FIFO (%)	C_EDD (%)
TH (lot)	2179	2183	2180	0.05	– 0.14
CT (h)	Seven hundred and seventy-eight; eat	783	777	0.13	0.77
VAR	481.97	478.99	484.24	– 0.47	– 1.10
HLCT (h)	903	906	902	0.11	0.44
HLVAR	791.74	814.54	818.00	– 3.32	– 0.42
CLCT (h)	768	772	766	0.26	0.78
CLVAR	452.48	447.72	452.59	– 0.02	– 1.09
ODR (%)	52.72	53.45	54.32	1.60	0.87
HLODR (%)	59.38	60.63	60.00	0.62	– 0.63

(1) No matter what dispatching rules are adopted, TH is almost unchanged under different dispatching rules, which is determined by the limited number of fixed WIP on the production line. Except for dispatch rules SPRT and FIFO, the performance indicators CT, HLCT, CLCT, VAR, HLVAR and CLVAR of RPELM_FS under other dispatch rules are slightly improved in some indicators compared with FIFO and EDD releasing rules, but the performance in some indicators is decreased instead.
(2) when the dispatch rule is SPRT, the important performance indexes of RPELM_FS, such as CT, ODR and HLODR, are obviously lower than FIFO and EDD, which indicates that RPELM is not suitable for semiconductor production lines with dispatch rule of SPRT.
(3) Under the condition that the dispatching rule is FIFO, all indexes of RPELM_FS are improved in different degrees compared with FIFO and EDD, especially RPELM can improve 1.03% and 6.67% respectively compared with ODR and HLODR of FIFO; Compared with EDD, HLODR can be improved by 2.86%.

(4) Compared with FIFO, RPELM_FS can improve ODR performance by 1.15% and HLODR performance by 3.08% when the dispatching rule is SPT. However, compared with EDD releasing strategy, it has no obvious advantages. Compared with FIFO and EDD, RPELM_FS can improve the performance index of HLODR by 2.81% and 2.28%, respectively, but it hardly improves other performance indexes.

(5) When the rules are SPRT and LS, compared with FIFO and EDD, RPELM_FS has almost no advantage in performance index, but has obvious decline in some performance indexes.

To sum up, when RPELM_FS is used together with dispatch rule FIFO, the performance index of semiconductor production system can be comprehensively improved. When the production line pays more attention to ODR or HLODR, RPELM_FS can also be used at the same time as the dispatching rules SPT and LPT. When the dispatching rule is LPTH or EDD, if the releasing rule is RPELM_FS, the production line will not be improved. It should be noted that the dispatching rule SPRT and the releasing rule RPELM_FS should not be adopted at the same time, because the overall effect obtained by adopting these two rules at the same time is very poor. The above results are caused by the sensitivity of extreme learning machine to data. Extreme learning machine is more effective for data generated by simulation when dispatching rule is FIFO, but less effective for data generated by simulation when other dispatching rules are adopted, especially for data generated by simulation using SPRT dispatching rule.

5.3.3.3 Experiments Under Different Emergency Workpiece Ratio

The different proportion of urgent parts in the order will affect the performance index of the production line, so the scheduling strategies on the production line, including the release control strategy and dispatching rules, should consider the influencing factor of the proportion of urgent parts in the order. To study whether RPELM_FS can still be applied to the semiconductor simulation model under orders with different emergency parts ratios, we selected four groups of orders with different emergency parts ratios and conducted simulation research on BL model. The simulation lasts for 300 days, in which the first 30 days are used as the preheating period. The simulation results are shown in Tables 5.30 and 5.31.

From Tables 5.30 and 5.31, the following conclusions can be drawn:

(1) The TH and CT performance indexes of FIFO and EDD releasing rules are the same. It can be seen that under the condition that the releasing rules are FIFO and EDD, the different proportions of urgent parts in the order will not affect the chip output and the average processing time, because the proportions of urgent parts will not affect the semiconductor scheduling under these two releasing rules. However, the releasing rule of RPELM_FS is related to the emergency workpiece ratio, so TH is different under the simulation of four groups of production lines with different emergency workpiece ratios.

Table 5.30 Simulation results of releasing strategy under different emergency workpiece proportions

	FIFO				EDD				RPELM_FS			
	TH (lot)	CT (h)	ODR (%)	HLODR (%)	TH (lot)	CT (h)	ODR (%)	HLODR (%)	TH (lot)	CT (h)	ODR (%)	HLODR (%)
One	2065	949	46.84	18.90	2072	948	48.53	23.17	2075	946	48.23	23.17
2	2065	949	47.39	26.75	2072	949	47.57	28.93	2073	945	47.70	28.96
Three	2065	949	46.98	25.56	2072	947	48.51	30.00	2069	947	49.72	32.22
Four	2065	949	42.65	20.00	2072	948	43.35	24.29	2075	946	43.14	24.29

Table 5.31 Comparison of simulation results of releasing strategies under different emergency workpiece proportions

	Improvement of RPELM_FS relative to FIFO performance				Improvement of performance of RPELM_FS relative to EDD			
	TH (lot) (%)	CT (h) (%)	ODR (%)	HLODR (%)	TH (lot) (%)	CT (h) (%)	ODR (%)	HLODR (%)
One	0.48	0.32	1.39	4.27	0.14	0.21	– 0.30	0.00
2	0.39	0.42	0.31	2.21	0.05	0.42	0.13	0.03
Three	0.19	0.21	2.74	6.66	– 0.14	0.00	1.21	2.22
Four	0.48	0.32	0.49	4.29	0.14	0.21	– 0.21	0.00

(2) In the first simulation, the ODR and HLODR of RPELM_FS are improved by 1.39% and 4.27% compared with FIFO, but they are not improved compared with EDD. The simulation results of Group 2 and Group 4 are the same as those of Group 1, except that the improvement of RPELM relative to FIFO is different, among which RPELM_FS has a maximum improvement of 1.39% in ODR and a maximum improvement of 4.29% in HLODR performance index.
(3) Under the third group of emergency parts ratio, ODR and HLODR, which are different from the other three groups of RPELM_FS, are improved compared with FIFO and EDD releasing rules, and the improvement ranges are 2.74% and 6.66% respectively compared with FIFO, and 1.2 1% and 2.2 2% respectively compared with EDD releasing strategy.

From the above conclusion, we can infer that RPELM_FS is still applicable under the condition that the proportion of urgent parts in different orders is different.

5.4 Summary

In this chapter, firstly, the attribute set related to workpiece releasing priority is analyzed, including workpiece attribute and processing area attribute in the whole production line, and then the selected real-time state is applied to the release control strategy of RPELM_FS and simulated on the MIMAC simulation model. The conclusion shows that the performance of RPELM_FS method after attribute selection can be improved compared with FIFO, EDD and RPELM. In addition, in order to further illustrate the correctness of RPELM_FS strategy and the practicability of attribute selection method, we simulated on BL model with different dispatching rules, and concluded that RPELM_FS is suitable for semiconductor production line models with FIFO, SPT and LPT dispatching rules. At last, the influence of the proportion of urgent parts on the releasing strategy of RPELM_FS is studied, and it is concluded that RPELM is also effective in ordering urgent parts with different proportions.

References

1. Liu W, Chua TJ, Cai TX (2005) Practical lot release methodology for semiconductor back-end manufacturing. Prod Plan Control 16(3):297–308
2. Wang KJ, Chiu CC, Gong DD (2010) Efficient job-release play development for semiconductor assembly and testing in GA. In: Conference on machine learning and cybernetic, pp 1205–1210
3. Qidi W, Fei Q, Li L, Zundan W (2006) Semiconductor manufacturing system scheduling. Electronic Industry Press, Beijing
4. You L, Zhibin J, Na L, Cheng L (2011) Overview of release control in wafer manufacturing system. Indust Eng Manage 16(6):108–114
5. Zhao Qi W, Zhiming. (2010) Research on Semiconductor Production Scheduling and Simulation: [Master Thesis]. Management Science and Engineering, Shanghai Jiaotong University, Shanghai
6. Spearman ML, Woodruff DL, Hoop WJ (1990) CONWIP: a pull alternative to kanban. Int J Prod Res 28(5):879–894
7. Lozinski C, Glassey CR (1989) Bottleneck starvation indicators for shop floor control. Semicond Manuf 1(1):36–46
8. Glassery CR, Resende MGC (1998) Close-loop job release control for VLSI circuits manufacturing. Semicond Manuf 1(4):147–153
9. Wein LM (1998) Scheduling semiconductor wafer fabrication. Semicond Manuf 1(2):115–130
10. Kim YD, Lee DH, Kim JU (1998) A Simulation study on lot release control, mask scheduling, and batch scheduling in semiconductor wafer fabrication facilities. J Manuf Syst 17(2):107–117
11. Li Y, Jiang ZB (2012) A pull VPLs based release policy and dispatching rule for semiconductor wafer fabrication. In: 8th IEEE International conference on automation science and engineering, pp 396–400
12. Kim J, Leachman RC, Suhn B (1996) Dynamic release control policy for the semiconductor wafer fabrication lines. J Oper Res Soc 47(12):1516–1525
13. Khaled S, Kilany E (2011) Wafer lot release policies based on the continuous and periodic review of WIP levels. In: IEEE International conference on industrial engineering and engineering management, pp 1700–1704
14. Chung SH, Lai CM (2006) Job releasing and throughput planning for wafer fabrication under demand fluctuating make-to-stock environment. Int J Adv Manuf Technol 31:316–327
15. Wu CC, Hsu PH, Lai KJ (2011) Simulated-annealing heuristics for the single-machine scheduling problem with learning and unequal job release times. J Manuf Syst 30:54–62
16. Adil B, Mustafa G (2011) A simulation based approach to analyses the effects of job release on the performance of a multi-stage job-shop with processing flexibilit y. Int J Prod Res 49(2):585–610
17. Chua TJ, Liu MW, Wang FY (2007) An intelligent multi-constraint finite capacity-based lot release system for semiconductor backend assembly environment. Rob Comput Integr Manuf 23:326–338
18. Wang KJ, Chiu CC, Gong DC, Hou TC (2011) An efficient job-releasing strategy for semiconductor turnkey factory. Prod Plan Control 22(7):660–675
19. Rezaie K, Eivazy H, Nazari-Shirkouhi S (2009) A novel release policy for hybrid make-to-stock/make-to-order semiconductor manufacturing systems. Comput Soc 443–447
20. Logy AEK, Khaled SEK, Aziz EES (2009) Modeling and simulation of re-entrant flow shop scheduling: an application in semiconductor manufacturing. In: IEEE International conference on automation science and engineering, pp 211–216
21. Jr AZ, Hodgson TJ, Weintraub AJ (2003) Integrated job release and shop-floor scheduling to minimize WIP and meet due-dates. Int J Prod Res 41(1):31–45
22. Wang ZJ, Chen J (2009) Release control for hot orders based on TOC theory for semiconductor manufacturing line. In: Proceedings of the 7th Asian control conference, pp 1154–1157
23. Kim YD, Kim JU, Lim SK (1998) Due-date based scheduling and control policies in a multiproduct semiconductor wafer fabrication facilities. IEEE Trans Semicond Manuf 11(1):155–164

24. Wang ZT, Qiao F, Wu OD (2005) A new compound priority control strategy in semiconductor wafer fabrication. IEEE Trans Semicond Manuf 80–83
25. Qi C, Appa IS, Stanley BG (2009) An efficient new job release control methodology. Int J Prod Res 47(3):703–731
26. Qi C, Appa IS, Stanley BG (2008) Impact of production control and system factors in semiconductor wafer fabrication. IEEE Trans Semiconduc Manuf 21(3):376–389
27. Qi C, Appa IS (2005) Job release based on WIPLOAD control in semiconductor wafer fabrication. In: Electronics packaging technology conference, pp 665–670
28. Bahaji N, Kuhl ME (2008) A simulation study of new multiobjective composite dispatching rules, CONWIP, and push lot release in semiconductor fabrication. Int J Prod Res 46(14):3801–3824(24)
29. Wang Z, Qidi W, Qiao F (2007) A lot dispatching strategy integrating WIP management and wafer start control. IEEE Trans Autom Sci Eng 4(4):579–583
30. Li M, Yang F, Uzsoy R, Jie X (2016) A metamodel-based Monte Carlo simulation approach for responsive production planning of manufacturing systems. J Manuf Syst 38:114–133
31. Chen ZB, Pan XW, Li L et al. (2014) A new release control policy (WRELM) for semiconductor wafer fabrication facilities. In: IEEE 11th International conference on networking, sensing and control (ICNSC), pp 64–68
32. Yao S, Jiang Z, Li N et al. (2010) A decentralized VPLs based control policy for semiconductor manufacturing. In: IEEE International conference on industrial engineering and engineering management, pp 1251–1255
33. Sun RJ, Wang ZJ (2008) DC-WIP —— a new release rule of multi-orders for semiconductor manufacturing lines. In: International conference on system simulation and semiconductor manufacture, pp 1395–1399

Chapter 6
Data-Driving Dynamic Scheduling of Semiconductor Manufacturing System

Since 1990s, with the development of manufacturing informatization, a large amount of data has been accumulated in the production process, and data mining has been applied in manufacturing industry. On the basis of traditional modeling and optimization methods of production scheduling problems, scholars at home and abroad extract key scheduling information hidden in a large number of data by means of feature analysis, data mining and simulation based on a large number of historical data, real-time data and relevant scheduling simulation data in the actual scheduling environment, and use the above information to establish a data-based production process-related scheduling model or dynamically determine the key parameters of the production process-related scheduling model. Data mining can obtain knowledge from relevant data to improve decision-making and increase production. At the same time, data visualization can give decision-makers a more intuitive understanding and help decision-makers better understand and use scheduling rules.

6.1 Dynamic Dispatching Rules

Dynamic Dispatching Rule (DDR) is the characteristic of group behavior which is optimized by indirect communication between individual ants based on pheromone in ant colony ecosystem. According to the equipment characteristics of semiconductor manufacturing production line, a dispatching rule which can consider the whole production line is obtained to realize adaptive dynamic scheduling of semiconductor production line.

L. Li et al., *Data-Driven Scheduling of Semiconductor Manufacturing Systems*,
Advanced and Intelligent Manufacturing in China,
https://doi.org/10.1007/978-981-19-7588-2_6

6.1.1 Definition of Parameters and Variables

Firstly, the parameters and variables of DDR are defined as follows:

i	Available equipment index number
id	Downstream equipment index number of equipment i
im	Menu index number of equipment i
iu	Upstream equipment index number of equipment i
k	Batch index number of queued workpiece group on batch processing equipment i
n	Index number of workpieces always queued in front of equipment i
t	Dispatch decision point, that is, dispatch time
v	Process menu index number of downstream equipment id
B_i	Processing capacity of batch processing equipment i
B_{id}	Processing capacity of downstream equipment id
D_n	Delivery time of workpiece n
F_n	The ratio of average processing cycle (sum of processing time and queuing time) to processing time of workpiece n
M_i	Number of process menus on equipment i
N_{id}	Number of workpieces queued in front of downstream equipment id
N_{im}	Number of workpieces queued in front of the equipment i to use the process menu im
P_i^n	Occupancy time of workpiece on equipment n i
P_{im}	Processing time of process menu mi on equipment i
P_{id}^n	Occupancy time of workpiece n on downstream equipment id
P_{id}^v	Processing time of process menu v on downstream equipment id
Q_i^n	Residence time of queued workpieces n on equipment i
R_i^n	Remaining processing time on workpiece n equipment i
S_n	Selection probability of workpiece n
T_{id}	Available time of downstream equipment id every day
Γ_k	Selection Probability of Workpiece Group Batch k
$\tau_i^n(t)$	The urgency of the equipment n to deal with the workpiece i at any time t
$\tau_{id}^n(t)$	The load degree of downstream equipment id that can complete the next working procedure of the workpiece n at the moment t
x_i^B	Binary variables. If the device is a bottleneck device i at the moment t, $x_i^B = 1$; Otherwise, $x_i^B = 0$
x_{id}^I	Binary variables. If the downstream equipment id is in idle state at the moment t, $x_{id}^I = 1$; Otherwise, $x_{id}^I = 0$
x_n^H	Binary variables. If the workpiece n is an emergency workpiece at any time t, $x_n^H = 1$; Otherwise $x_n^H = 0$
x_n^{im}	Binary variables. If the workpiece n adopts the process menu m on the equipment i, $x_n^{im} = 1$; Otherwise, $x_n^{im} = 0$

(continued)

(continued)

$x^{id}_{n,im}$	Binary variables. If the downstream equipment id in the next process of processing the workpiece n is in an idle state at the moment t, and the menu im is used in the equipment instead of the workpiece, $x^{id}_{n,im} = 1$; Otherwise, $x^{id}_{n,im} = 0$

6.1.2 Hypothesis

DDR makes the following assumptions in solving the dispatching problem:

(1) the information related to dispatching is known, such as workpiece processing time, WIP number in front of equipment, equipment available time, etc. These data can be obtained by MES or other automation systems of enterprises;

(2) The dispatching decision for non-batch processing equipment mainly focuses on the on-time delivery rate of workpieces and the rapid movement of WIP on the production line;

(3) There are two steps for dispatching decision of batch processing equipment:

① There are two main constraints in batch processing of workpieces, namely, only workpieces using the same process menu on the equipment can be batch processed, and the number of batch processed workpieces cannot exceed the maximum processing batch of the equipment. In addition, the capacity utilization rate and time utilization rate of batch processing equipment are considered in a compromise;

② To determine the priority of batch workpieces, at this time, its focus is the same as that of non-batch processing equipment, which is also the on-time delivery rate of workpieces and the rapid movement of WIP in the production line.

(4) The processing time of each batch of workpieces has nothing to do with the number of workpieces that make up the batch;

(5) Once the equipment starts processing a certain batch of workpieces, it cannot add or remove workpieces from the batch, and the equipment will keep the processing state until the workpiece processing is completed.

6.1.3 Decision-Making Process

The decision flow of DDR algorithm is shown in Fig. 6.1, and the specific steps are as follows:

Step 1: When the equipment becomes available at the moment, determine whether the equipment is batch processing equipment. If yes, go to step 2; Otherwise, go to Step 6.

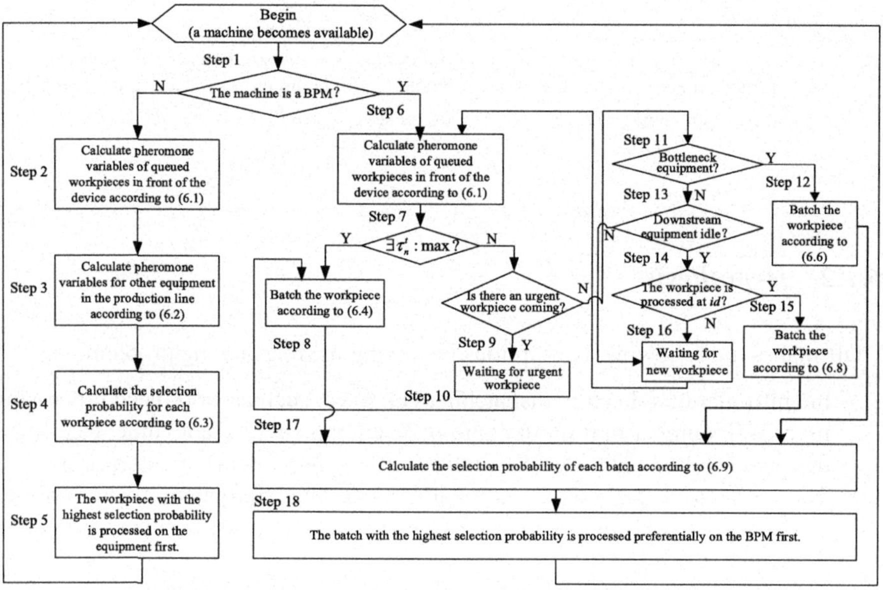

Fig. 6.1 Decision flow of DDR

Step 2: Calculate the information variables of queued workpieces before the equipment i

$$\tau_i^n(t) = \begin{cases} \text{MAX} & R_i^n \times F_n \geq D_n - t \\ \frac{R_i^n \times F_n}{(D_n - t + 1)} - \frac{P_i^n}{\sum_n P_i^n} & R_i^n \times F_n < D_n - t \end{cases} \tag{6.1}$$

Formula (6.1) is designed to meet the requirements of customers for on-time delivery. At the moment t, the greater the ratio of the theoretical remaining processing time to the actual remaining processing time of each WIP, the tighter the delivery time. Accordingly, the higher the information variable value of the WIP, the easier it is to be selected by the equipment for priority processing. However, if the theoretical remaining processing time of the WIP is longer than the actual remaining processing time, which means that the WIP is very likely to be delayed, it will become an urgent workpiece, that is, it has the highest processing priority (MAX) on any equipment. In addition, the occupation time of each WIP will also affect its information variable value. The shorter the occupation time, the higher the information variable value, which can speed up the movement of WIP on equipment and improve the utilization rate of equipment.

Step 3: Calculate the information variables of other equipment on the production line

$$\tau_{id}^n(t) = \frac{\sum P_{id}^n}{T_{id}} \tag{6.2}$$

Equation (6.2) means that at moment t the heavier the equipment load, the higher the information variable. Obviously, when $\tau_i^n(t) \geq 1$, it means that the load of the equipment has exceeded its available time in one day, that is, it is considered that the equipment is in a bottleneck state. It is worth noting that there may be several equipment in the semiconductor production line that can complete the specific process of WIP. In this case, the significance of T_{id} is the available processing time of a class of equipment that can complete the process of WIP to be processed in one day.

Step 4: Calculate the selection probability of each queued workpiece

$$S_n = \begin{cases} Q_i^n & \tau_i^n(t) = \text{MAX} \\ \alpha_1 \tau_i^n(t) - \beta_1 \tau_{id}^n(t) & \tau_i^n(t) \neq \text{MAX} \end{cases} \tag{6.3}$$

Equation (6.3) means that at moment t, when solving the problem of WIP competing for equipment resources, the delivery time and occupation degree of WIP and the load condition of downstream equipment will be considered at the same time to ensure the rapid flow and on-time delivery rate of WIP.

Step 5: Select the workpiece with the highest selection probability to start processing on the equipment i.

Step 6: Use Formula (6.1) to calculate the information variables of queued workpieces before equipment i.

Step 7: Determine whether there are urgent workpieces (i.e., urgent workpieces) in the queue before the equipment, and if so, go to Step 8; Otherwise, go to Step 9.

Step 8: Batch workpieces according to Formula (6.4)

$$for\ im = 1 \text{ to } M_i$$

$$if\ 0 \leq \sum x_n^{im} < B_i$$

$$then\ Select\left\{\min\left\{\left(B_i - \sum x_n^{im}\right),\ \left(N_{im} - \sum x_n^{im}\right)\right\}\right\}\Big|_{\max(Q_i^n)} \tag{6.4}$$

$$else\ if\ \sum x_n^{im} \geq B_i$$

$$then\ Select\{B_i\}|_{\max((R_i^n \times F_n)-(D_n - t))}$$

Formula (6.4) means: for each process menu im of equipment, i if the number of emergency workpieces is less than B_i, check whether the common workpieces before queuing equipment i adopt the same menu as the emergency workpieces. If the number of ordinary workpieces that meet the conditions is small, select the batch of workpieces $B_i - \sum x_n^{im}$ before the equipment according to the principle that the longer the waiting time of workpieces, the higher the priority; Otherwise,

select all ordinary workpieces that meet the requirements ($N_{im} - \sum x_n^{im}$). If the number of urgent work pieces is greater than or equal to B_i, directly select the most urgent work pieces that meet the maximum processing batch and batch them. Then turn to step 17 to determine the processing priority of the batch workpieces.
Step 9: According to the Formula (6.1), judge whether the workpiece being processed or just finished processing on the upstream equipment iu of the batch processing equipment i is an emergency workpiece, and if there is an emergency workpiece, go to Step 10; Otherwise, go to Step 11.
Step 10: Wait for the arrival of urgent workpieces, and then turn to Step 8 to batch workpieces according to Formula (6.4).
Step 11: Determine whether the equipment is a bottleneck equipment according to Formula (6.5). If yes, go to step 12; Otherwise, go to Step 13 i.

$$If \sum_{im} N_{im} \geq (24B_i/\min(P_{im})),\ then\ x_i^B = 1 \tag{6.5}$$

Formula (6.5) means that if the queued workpieces in the buffer zone of batch processing equipment i have exceeded their daily maximum processing capacity (i.e., the maximum workpieces that can be processed within 24 h), the equipment is considered to be in a bottleneck state.
Step 12: Batch workpieces according to Formula (6.6), and then turn to Step 17 to determine the processing priority of the batch workpieces.

$$Select\{B_i\}|_{\max(Q_i^n)} \tag{6.6}$$

Formula (6.6) means that batch processing is carried out according to the process menu im of batch processing equipment i used by queued workpieces. If the workpieces using the same process menu exceed the maximum processing batch, batch processing is carried out according to the principle that the longer the waiting time of workpieces, the higher the priority.
Step 13: Determine whether the downstream equipment id is idle equipment according to Formula (6.7). If yes, go to step 14; Otherwise, go to Step 16.

$$If \sum_{im} N_{id} \geq \left(24B_i/\min\left(P_{id}^v\right)\right),\ then\ x_{id}^I = 1 \tag{6.7}$$

Equation (6.7) means that if the queued workpieces in the buffer zone of downstream equipment id are lower than their daily minimum processing capacity (i.e., the minimum workpieces that can be processed within 24 h), the equipment is considered to be in idle state.
Step 14: Judging whether there are workpieces waiting to be processed in idle downstream equipment id in the next working procedure in the queued workpieces of the equipment i; If so, go to step 15; Otherwise, go to Step 16.
Step 15: Batch the workpieces according to Formula (6.8).

$$for\ im = 1 \text{ to } M_i$$

$$if\ 0 \le \sum x_{n,\ im}^{id} < B_i$$

$$then\ \ Select\left\{\min\left[\left(B_i - \sum x_{n,\ im}^{id}\right), \left(N_{im} - \sum x_{n,\ im}^{id}\right)\right]\right\}\Big|_{\max(Q_i^n)} \quad (6.8)$$

$$else\ if\ \sum x_{n,\ im}^{id} \ge B_i$$

$$then\ \ Select\{B_i\}|_{\max(Q_i^n)}$$

Formula (6.8) means: For each process menu im of equipment i, check the number of workpieces to be processed on idle equipment in the next process; if it is less than the maximum processing batch of equipment B_i, check whether there are other workpieces that use the same process menu as these workpieces; if there are more qualified workpieces, select several non-emergency workpieces according to the principle that the longer the waiting time of workpieces, the more priority is given to meet the maximum processing batch; If it is greater than or equal to the maximum processing batch, directly select the workpiece with the longest queuing time and meet the maximum processing batch. Then turn to Step 17 to determine the processing priority of batch workpieces.

Step 16: Wait for the arrival of the new workpiece and turn to Step 6 to restart the dispatching decision.

Step 17: Determine the priority of each batch of workpieces according to Formula (6.9).

$$\Gamma_k = \alpha_2 \frac{N_{ik}^h}{B_i} + \beta_2 \frac{B_k}{\max(B_k)} - \gamma \frac{P_i^k}{\max\left(P_i^k\right)} - \sigma\left(N_{id}^k / \left(\sum_k N_{id}^k + 1\right)\right) \quad (6.9)$$

where N_{ik}^h is the number of urgent workpieces in the batch k; B_k is the batch size of the batch k; P_i^k is the occupation time of the batch k on the equipment i; N_{id}^k is the maximum load of downstream equipment for batch processing; $(\alpha_2, \beta_2, \gamma, \sigma)$ is an index to measure the relative importance of these four items.

The first term of Formula (6.9) is the proportion of urgent workpieces in the batch k processing, which corresponds to the on-time delivery rate index; The second item is the ratio of the batch processing batch n to the maximum batch processing in all batches, which corresponds to the processing cycle, moving steps and equipment utilization index; The third item is the ratio of batch n processing time to the maximum processing time in all batches, which corresponds to the occupation time of the workpiece to the equipment, which is related to the processing cycle index and can also reflect the moving step index; The fourth item is the load degree of downstream equipment, which is related to the equipment utilization index and

can also reflect the moving steps index. Therefore, with the different concerned indexes or the change of manufacturing environment, the desired performance indexes can be obtained by adjusting the corresponding parameters $(\alpha_2, \beta_2, \gamma, \sigma)$.
Step 18: Select the batch workpieces with the highest selection probability to start processing on the equipment i.

6.1.4 *Simulation and Verification*

Taking the historical data of many production lines of 6-inch silicon wafers in a semiconductor manufacturing enterprise in Shanghai as the research object, according to the actual needs of the enterprise, combined with the dynamic modeling method, a production line simulation model consistent with the actual production line was built by Siemens Tecnomatix Plant Simulation software as the research platform for simulation verification.

At present, there are nine processing areas in the production line of this enterprise, which are: injection area, lithography area, sputtering area, diffusion area, dry etching area, wet etching area, back thinning area, PVM test area and BMMSTOK microscopic inspection area. The dispatching rules used are based on manual priority scheduling method, referred to as PRIOR for short. The main idea is to set the priority according to the manual experience, to ensure the products can be delivered on time to the greatest extent, that is, to meet the delivery time index.

It can be seen from Fig. 6.1 that in DDR, the information related to scheduling is encapsulated in the algorithm, and then weighted. by adjusting these weights $(\alpha_1, \beta_1, \alpha_2, \beta_2, \gamma, \sigma)$, the universality of DDR to the changing environment is realized. This means that when the value of $(\alpha_1, \beta_1, \alpha_2, \beta_2, \gamma, \sigma)$ is different, the performance indexes are different.

Assume that the six weighted parameter values $(\alpha_1, \beta_1, \alpha_2, \beta_2, \gamma, \sigma)$ of DDR are:$\alpha_1 = 0.5, \beta_1 = 0.5, \alpha_2 = 0.25, \beta_2 = 0.25, \gamma = 0.25, \sigma = 0.25$.

The following three cases are designed and verified by simulation for up to 3 months:

Case 1: Adopt the PRIOR rule of the enterprise.

Case 2: For all equipment without special restrictions in the production line, its scheduling rules are replaced by DDR.

Case 3: Only equipment in the production line with no special restrictions and daily equipment utilization rate greater than 60% is replaced by DDR, and other equipment still adopts PRIOR rule.

From the short-term performance index: average daily Move and average daily equipment Utility, and the long-term performance index: Throughput, average processing Cycle Time, CT), ideal processing time/real processing time (IPT/RPT) to compare the optimization results of Case 1, Case 2 and Case 3 on the production line.

As the Move value is of order of magnitude 10^3, the output quantity is of order of magnitude $10^1 \sim 10^2$, the average processing time is of order of magnitude 10^1,

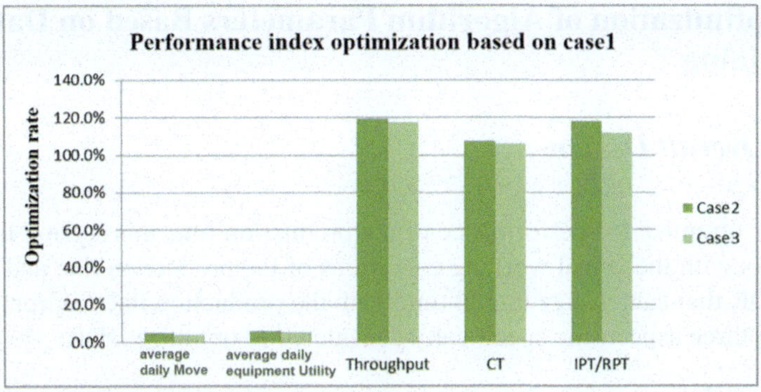

Fig. 6.2 Comparison of performance index results

and the Utility value and ideal processing time/actual processing time are of order of magnitude 10^{-1}, this paper takes the value of Case 1 as the basic value, and sets it as 1, with Case 2 and Case 3 indicating the improvement degree of Case 1.

The experimental results are shown in Fig. 6.2, in which Throughput indicates the number of tablets; CT shows Cycle Time, that is, processing time; IPT/RPT represents ideal processing time/real processing time, which is the ratio of ideal processing time to actual processing time.

DDR is superior to PRIOR rule in both short-term performance index and long-term performance index, especially in long-term performance index, which is improved by 100% based on PRIOR rule. However, adopting DDR for all devices has little improvement compared with adopting DDR for bottleneck devices only. This is because the non-bottleneck equipment has sufficient resources, and when the workpiece arrives, it can be processed immediately, instead of waiting in the buffer, so it only needs to adopt simple FIFO rules.

It can be seen from Table 6.1 that the simulation runs for 90 days, and all devices adopt DDR, which is 7.5 times of the simulation time when only bottleneck devices adopt DDR, but the optimization effect is not obvious. This is because DDR calculation is complicated, especially the time complexity of Step 9 is high. Therefore, in the subsequent experiments, this paper only calls DDR to simulate the equipment with equipment utilization rate greater than 60%.

Table 6.1 Simulation time comparison

Case	Time consumption (h)
Case 1	Three
Case 2	45
Case 3	Six

6.2 Optimization of Algorithm Parameters Based on Data Mining

6.2.1 Overall Design

To further optimize the performance of the production line, this section links DDR parameters with the actual working conditions of the production line and improves it into ADR that can change in real time with the production line environment. The following three aspects are specifically considered to optimize DDR:

(1) Load state

Under different load conditions, better values of Move and Utility can be obtained by adjusting algorithm parameters $(\alpha_1, \beta_1, \alpha_2, \beta_2, \gamma, \sigma)$. Here, we first define two concepts related to load state: Required Capacity, RC) and Available Capacity, AC).

Demand capacity refers to the processing time required for the workpiece to be processed in the buffer zone of processing equipment. For non-batch processing equipment, the required capacity refers to the total processing time of all the workpieces waiting to be processed before the equipment; However, for batch processing equipment, the required production capacity is not simply added up, but first batch according to the equipment processing menu, and then add up each processing time. In the actual semiconductor production process, the equipment menu in the same processing area is interchangeable, so the required capacity can also be extended to the interchangeable equipment group in the processing area.

Available capacity refers to the available processing time of processing equipment. From the point of view of Overall Equipment Efficiency, OEE), besides downtime and preventive maintenance time have certain influence on production capacity, abnormal processing such as processing time of engineering card and file control sheet should be deducted from production capacity. In addition, to maintain the stability of the schedule, some protective capacity must be reserved. Therefore, the available capacity of a certain equipment in one day can be expressed by Formula (6.10), and the unit is minutes.

$$AC = (1 - DT - PM - EG - MD - PC) \times 1440 \tag{6.10}$$

where DT is the Down Time ratio; PM is the proportion of preventive maintenance time; EG is the time proportion of processing engineering cards; MD is the time proportion for processing the document control sheet; PC is the time ratio of protective capacity.

The Load can be expressed by $RC/AC \times 100\%$. The load state of the production line can be obtained by weighting and averaging each available equipment or interchangeable equipment group in the processing area.

When $line\ load > 100\%$, it is considered that the current production line is in an overload state; When $line\ load = 100\%$, it is considered that the current production line is in full load state; When $90\% < line\ load < 100\%$, it is considered that the current production line is under heavy load; When $75\% < line\ load < 90\%$, it is considered that the current production line is under-loaded; When $line\ load < 75\%$, it is considered that the current production line is in a light load state.

(2) Real-time status related to production line performance

The research background is the semiconductor production line of an enterprise. The short-term performance indicators of the line are Move and Utility, and the two real-time states that are most closely related to them are the emergency workpiece ratio and the post-1/3 lithography workpiece ratio.

Emergency workpiece is a special existence in the production line system because any equipment is idle, when selecting workpieces from the buffer zone, if there are emergency workpieces in the buffer zone, they must be processed first, which means that emergency workpieces will have certain interference and influence on the normal processing sequence of workpieces. Therefore, the proportion of emergency parts r_h in production line is an important index to control the stable operation of production line system. The smaller the proportion is, the less interference will be caused to the production line system, thus making the production line system more controllable to a certain extent.

The semiconductor mentioned here usually refers to an electronic chip, which is an integrated circuit composed of electrical connections layer by layer. And each layer needs lithography, so lithography is the key index to measure the completion stage of semiconductor products. In this paper, the real-time state of the post-1/3 lithography workpiece ratio (i.e. r_p, the remaining 1/3 lithography process) is used as an important index to measure the number of wafers, and the higher the ratio, the greater the amount of WIP to be released, thus relieving the equipment pressure to a certain extent.

Establish a logical relationship between the values $(\alpha_1, \beta_1, \alpha_2, \beta_2, \gamma, \sigma)$ and (r_h, r_p), and express it as:

$$\begin{aligned}
\alpha_1 &= a_1 \cdot r_h + b_1 \cdot r_p + c_1 \\
\beta_1 &= a_2 \cdot r_h + b_2 \cdot r_p + c_2 \\
\alpha_2 &= a_3 \cdot r_h + b_3 \cdot r_p + c_3 \\
\beta_2 &= a_4 \cdot r_h + b_4 \cdot r_p + c_4 \\
\gamma &= a_5 \cdot r_h + b_5 \cdot r_p + c_5 \\
\sigma &= a_6 \cdot r_h + b_6 \cdot r_p + c_6
\end{aligned} \tag{6.11}$$

Therefore, the best value $(\alpha_1, \beta_1, \alpha_2, \beta_2, \gamma, \sigma)$ can be obtained as long as the appropriate $(a_i, b_i, c_i, i \in \{1, \ldots, 6\})$ is selected, so as to achieve the purpose of optimizing Move and Utility.

(3) Considering the real-time state related to the lithography area separately

Considering that r_h, r_p is the real-time state information of the whole production line system, which can not reflect the actual state of each processing area, and $(\alpha_1, \beta_1, \alpha_2, \beta_2, \gamma, \sigma)$ may have adverse effects on the setting of some processing areas according to the production line state, so we try to consider the emergency workpiece ratio and the last 1/3 photoetching workpiece ratio of each processing area independently. Here, priority is given to the bottleneck processing area-lithography area.

The emergency workpiece ratio and the post-1/3 photoetching workpiece ratio in the photoetching area are separately recorded as (r_h_photo, r_p_photo), which is shown as follows:

$$\begin{aligned} \alpha_1 &= a_1 \cdot r_h\_photo + b_1 \cdot r_p\_photo + c_1 \\ \beta_1 &= a_2 \cdot r_h\_photo + b_2 \cdot r_p\_photo + c_2 \\ \alpha_2 &= a_3 \cdot r_h + b_3 \cdot r_p + c_3 \\ \beta_2 &= a_4 \cdot r_h + b_4 \cdot r_p + c_4 \\ \gamma &= a_5 \cdot r_h + b_5 \cdot r_p + c_5 \\ \sigma &= a_6 \cdot r_h + b_6 \cdot r_p + c_6 \end{aligned} \tag{6.12}$$

6.2.2 *Algorithm Design*

6.2.2.1 BP Neural Network

Back propagation (BP) network is a multi-layer feedforward neural network based on error back propagation (BP algorithm), which was put forward by scientists led by Rumelhart and McCelland in 1986. It is one of the most widely used neural networks at present.

A trained BP network can theoretically realize any nonlinear mapping between input and output and can approach any nonlinear function. Therefore, BP network has strong fault tolerance, self-learning habit and adaptability [1].

Figure 6.3 shows the typical topology of three-layer BP network, which is composed of input layer, hidden layer, and output layer, and also includes the transfer function and training function among the layers. BP network flows through unidirectional connection from the input layer to the output layer, only the neurons between the front and back adjacent layers are fully connected with each other, and the signals received from the upper layer are transmitted to the next layer of neurons, but there is no connection between the neurons in the same layer and no feedback between the neurons [2].

Take the simulation experiment of DDR parameter optimization as an example.

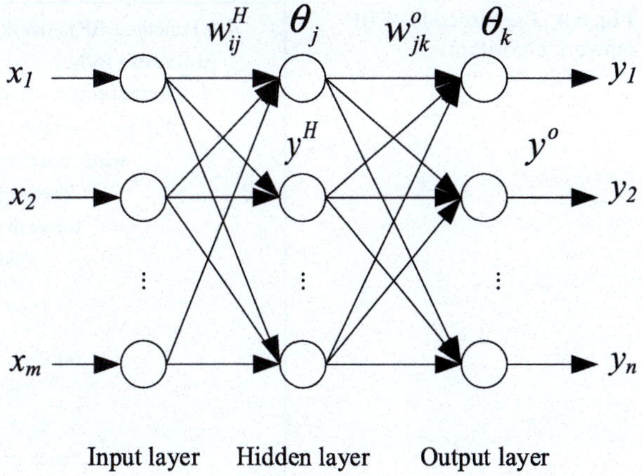

Fig. 6.3 Topology structure of three-tier BP network

The value $(\alpha_1, \beta_1, \alpha_2, \beta_2, \gamma, \sigma)$ is taken as six nodes of the input layer of BP network, and the optimized production line performance indexes (Move and Utility) are taken as two nodes of the output layer of BP network. Among them, the number of hidden layer nodes has a certain influence on BP network training. If there are too many hidden layer nodes, the training time is too long; However, if the number of nodes in the hidden layer is too small, the fault tolerance, generalization ability and recognition ability of unlearned test samples will be poor. Therefore, refer to the formula $h = \sqrt{m + n} + a$ of the number of nodes in the hidden layer (which a is a constant between 1 and 10), and set the number of nodes in the hidden layer to 5.

For the transfer function of BP neural network, differentiable monotonically increasing functions are usually selected, such as linear function, logarithmic S-type function, and tangent S-type function. Therefore, S-type function is selected in this chapter, so that the output of the whole BP network will be limited to a very small range.

According to different applications, BP network provides a variety of training functions. For the function approximation network, the training function trainlm has the fastest convergence speed and small convergence error; For pattern recognition networks, the training function trainrp converges fastest. Trainscg, the training function of the variable gradient algorithm, has good performance when the network scale is relatively large. In this paper, the problem of function fitting approximation is solved, so trainlm is used as the training function of BP network.

As shown in Fig. 6.4, the training process of BP network algorithm can be summarized as the following steps:

Step 1: Initialize network weights

The weight of network connection between every two neurons ω_{ij} is initialized to a small random number (for example $-1.0 \sim 1.0 - 0.5 \sim 0.5$, or according to the problem itself). At the same time, each neuron has an offset θ_i and is also initialized to a random number.

Fig. 6.4 Pseudo-code of BP network algorithm

```
// Function: BP network training process pseudo-code
procedure BPNN
    Initialization, include the ω_ij and θ_i
    for each sample X
        while not stop
            //forwards propagation of the input
            for each unit j in the hidden and output layer
                //calculate the output O_j;
                O_j = 1/(1+e^(-S_j)) = 1/(1+e^(-(Σ_i ω_ij O_i + θ_j)))
            for each unit j in the output layer
                //calculate the error E_j;
                E_j = O_j(1 − O_j)(T_j − O_j);
            //back propagation of the error
            for each unit j in the hidden layer
                //calculate the error E_j;
                E_j = O_j(1 − O_j) Σ_k ω_jk E_k
            //adjust the network parameters
            for each network weight ω_ij
                ω_ij = ω_ij + (l)O_i E_j;
            for each biased θ_j
                θ_j = θ_j + (l)E_j;
        end of while not stop
    end of for each sample X
end of procedure
```

For each input sample x, process it according to Step 2.

Step 2: Forward propagation of input (feedforward network)

Firstly, the input layer of the network is provided according to the training samples x, and the output of each neuron is calculated. The calculation method of each neuron is the same, which is obtained by the linear combination of its inputs. The specific formula is

$$O_j = \frac{1}{1+e^{-S_j}} = \frac{1}{1+e^{-(\sum_i \omega_{ij} O_i + \theta_j)}}$$

where ω_{ij} is the network weight from the unit of the upper layer i to the unit j; O_i is the output of the unit at the upper level; As the bias of this cell, θ_j can be used as a threshold to change the activity of the cell. It can be seen from the above formula that the output of neuron depends on its total input $S_j = \sum_i \omega_{ij} O_i + \theta_j$, and then the final output is obtained by activation function $O_j = \frac{1}{1+e^{-S_j}}$, which becomes logistic function or sigmoid function, and can map the larger input value to a value between $0 \sim 1$ intervals. Because this function is nonlinear and differentiable, it also makes BP network algorithm model the linear indivisible classification problem, which greatly expands its application range.

Step 3: Reverse error propagation

Go all the way from Step 2, and finally get the actual output at the output layer. The error of each output unit j can be obtained by comparing with the expected output, as shown in the formula $E_j = O_j(1 - O_j)(T_j - O_j)$, where T_j is the expected output of the output unit j. The obtained error needs to propagate from the back to the front, and the error of the previous layer unit j can be calculated by the error of all the units k in the later layer connected with it. The specific formula is

$$E_j = O_j(1 - O_j)\sum_k \omega_{jk}E_k$$

The error of each neuron from the last hidden layer to the first hidden layer is obtained in turn.

Step 4: Network weight and neuron bias adjustment

In the process of processing, we can adjust the network weights and neuron thresholds while propagating errors backward. However, for convenience, the errors of all neurons are calculated first, and then the network weights and neuron thresholds are adjusted uniformly.

The method of weight adjustment is to start from the connection weight of the input layer and the first hidden layer, and proceed backward in turn, and each connection weight ω_{ij} is adjusted according to the formula $\omega_{ij} = \omega_{ij} + \Delta\omega_{ij} = \omega_{ij} + (l)O_i E_j$.

The adjustment method of neuron bias is to update each neuron j as shown in the formula $\theta_j = \theta_j + \Delta\theta_j = \theta_j + (l)E_j$.

Where l is the learning rate, which is usually a constant between 0 and 1. This parameter will also affect the performance of the algorithm. Experience shows that too small a learning rate will lead to slow learning, while too large a learning rate may cause the algorithm to vibrate between inappropriate solutions. An empirical rule is to set the learning rate as the reciprocal of iteration times t, that is $^1/_t$.

Step 5: End of judgment

For each sample, if the final output error is less than the acceptable range or the number of iterations t reaches a certain threshold, select the next sample and go to Step 2 to continue execution again; Otherwise, the number of iterations t is increased by 1, and then Step 2 is turned to continue to use the current sample for training.

In this paper, BP network algorithm will be used to train sample data, and with its excellent prediction ability, better parameters of dynamic dispatching rules will be obtained to improve Move and Utility.

6.2.2.2 Particle Swarm Optimization Algorithm

Particle Swarm Optimization (PSO) is an evolutionary computing technology developed by Kennedy and Eberhart in 1995, which comes from the simulation of a simplified social model [3].

Particle Swarm Optimization (PSO) operates according to the fitness value of particles. Each particle flies at a certain speed in the N-dimensional search space, and the flight speed is dynamically adjusted by the individual flight experience and the group flight experience.

Let the size of population be $pop\ Size$; $x_i = (x_{i1}, x_{i2}, \ldots\ldots, x_{in})$ is the current position of particles i; $v_i = (v_{i1}, v_{i2}, \ldots\ldots, v_{in})$ is the current velocity of the particle i; $P_i = (P_{i1}, P_{i2}, \ldots\ldots, P_{in})$ is the best position of the particle i in history, which is called the best position of the individual, that is, the best solution found by the particle itself. For the minimization problem, the best position in history refers to the position with the smallest fitness function value. The population is expressed as the objective function min $f(X)$. In the t-th generation, the update formula of the best position of the individual is expressed as:

$$P_i(t+1) = \begin{cases} P_i(t), iff(x_i(t+1)) \geq f(P_i(t)) \\ x_i(t+1), iff(x_i(t+1)) < f(P_i(t)) \end{cases} \tag{6.13}$$

The best position experienced by all particles in the population $P_g(t)$ is called the global best position, that is, the optimal solution found by the whole population at present. Then

$$\begin{aligned} &P_g(t) \in \{P_0(t), P_1(t), \ldots\ldots, P_s(t)\} | f\left(P_g(t)\right) = \\ &\min\{f(P_0(t)), f(P_1(t)), \ldots\ldots, f(P_s(t))\} \end{aligned} \tag{6.14}$$

The basic particle swarm optimization algorithm can update its own speed and new position according to the following evolutionary formula:

$$v_{ij}(t+1) = v_{ij}(t) + c_1 r_1(t)\left(P_{ij}(t) - x_{ij}(t)\right) + c_2 r_2(t)\left(P_{gj}(t) - x_{ij}(t)\right) \tag{6.15}$$

$$x_{ij}(t+1) = x_{ij}(t) + v_{ij}(t+1) \tag{6.16}$$

where, $v_{ij}(t), v_{ij}(t+1), x_{ij}(t), x_{ij}(t+1)$ represent the flight speed and position of the i-th particle in the t-th generation and the t + 1-th generation, respectively. Subscript i indicates particles i, j: the j-th dimension of velocity (or position), t: the t-th generation, and c_1, c_2 : learning factors, which are the acceleration constants of individual particles and group particles, respectively. Usually,$c_1, c_2 \in [0, 2]$; And r_1, r_2 are random numbers between [0,1].

It can be seen from the speed update formula in Formula (6.15) that c_1 refers to the step size of particles in the historical optimal direction and c_2 refers to the step size of particles in the global optimal direction are adjusted. In order to reduce the possibility of particles leaving the search space during evolution, usually, $v_{ij} \in [-v_{\max}, v_{\max}]$.

As shown in Fig. 6.5, the particle swarm optimization algorithm can be summarized as follows:

Step 1: Initialize population, position and speed;
Step 2: Calculate the fitness value of each particle;
Step 3: Compare the fitness value of each particle with the historical best position P_i that the particle has experienced, and if the current fitness value is better than P_i, take it as the historical best position;
Step 4: Compare the historical best position P_i of each particle with the global best position P_g, and if it is better, take it as the global best position P_g;

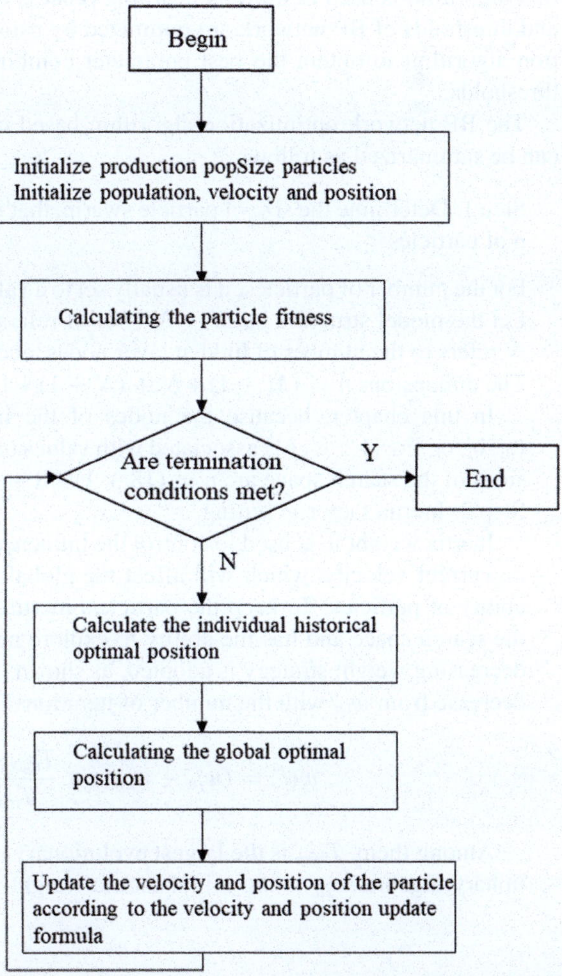

Fig. 6.5 Flow chart of particle swarm optimization

Step 5: Update the velocity and position of particles by using Formulas (6.15) and (6.16);
Step 6: Judge whether the termination condition of the algorithm is satisfied, if yes, the algorithm ends, otherwise, go to Step 2.

In this paper, Particle Swarm Optimization (PSO) is used to optimize the weights and thresholds of BP network algorithm, and further optimize the parameters of dynamic dispatching rules.

6.2.2.3 BP Network Optimization Algorithm Based on PSO

In view of the inherent problems of slow learning speed, easy to fall into local minimum and "over-learning" of BP neural network, the particle swarm optimization algorithm is used to train the learning process of BP network, and the weights and thresholds of BP network are optimized by using the particle swarm optimization algorithm to obtain the best parameter combination of network weights and thresholds.

The BP network optimization algorithm based on particle swarm optimization can be summarized as follows:

Step 1: Determine the size of particle swarm, that is, the number m and dimension n of particles

For the number of particles, it is usually set to a value between 10 and 40, $m = 10$. Let the model structure be $M - N - 1$: M refers to the number of input nodes, N refers to the number of hidden layer nodes, and 1 the number of output nodes. The dimensions $n = (M + 1) * N + (N + 1) * 1$ of the space are searched.

In this chapter, because the nodes of the input layer are 18 coefficients ($a_i, b_i, c_i, i \in \{1, \ldots, 6\}$) associated with values ($\alpha_1, \beta_1, \alpha_2, \beta_2, \gamma, \sigma$), the dimensions of the search space are $n = (18 + 1) * 8 + (8 + 1) * 1 = 161$.
Step 2: Inertia factor w setting.

Inertia weight w is used to control the influence of particle's previous velocity on current velocity, which will affect the global search ability and local search ability of particles. To keep the particle's inertia, it has the tendency to expand the search space and has the ability to explore new areas. In this paper, a linear decreasing weight strategy is adopted, as shown in Formula (6.16), which w can decrease from w_{in} with the number of iterations to w_{end}.

$$w(t) = (w_{in} - w_{end}) \times \frac{T_{\max} - t}{T_{\max}} + w_{end} \tag{6.17}$$

Among them, $T_{\max}$ is the largest evolutionary algebra and t:the current evolutionary algebra..w_{in} is the initial inertia weight and w_{end} is the inertia weight when iterating to the maximum algebra. Parameter is set to, $w_{in} = 0.9$, $w_{end} = 0.3$, $T_{\max} = 200$.

Step 3: Learning factors c_1 and c_2 settings.

c_1 and c_2 the weight of the statistical acceleration term representing the push to the sum position P_{ij}, P_{gj} of each particle, which are parameters for adjusting the role of the particle's own experience and social group experience in the whole cruise process. And c_1, c_2 are fixed constants, which c_1, c_2 are generally limited and equal and have a value range of [0,4]. In this chapter, $c_1 = 2$, $c_2 = 1.8$.

Step 4: Determine the fitness function

The training mean square error function E is used as the fitness evaluation function of particles to promote the search of population. The fitness function of particles is calculated according to Formula (6.18).

$$fitness = E = \frac{1}{N}\sum_{i=1}^{N}\left(y_{i(real)} - y_i\right)^2 \tag{6.18}$$

where N is the number of training samples, $y_{i(real)}$ is the actual value of sample i, and y_i is the model output value of sample i. Therefore, when the iteration of the algorithm stops, the position corresponding to the particle with the lowest fitness is the optimal solution of the problem.

Step 5: Initialize speed and position

Individuals are randomly generated, and each individual consists of two parts, the first part is the velocity matrix of particles, and the second part represents the position matrix of particles.m

In this paper, the weights and thresholds obtained from the initialization of BP network are taken as the initial positions of each particle in PSO algorithm.

Step 6: Evaluation

According to Formula (6.18), calculate the fitness of particles in the population under the BP neural network training samples.

Step 7: Extreme value update

Compare the current individual fitness value of the population with the individual fitness value before iteration. If the current value is better, make the current value replace the value before iteration, and save the current position as its individual extremum; otherwise, its individual extremum is the extremum of the previous generation. For global extremum, if the current fitness value of a particle in the existing population is better than the global historical optimal fitness value, let the current fitness value of the particle be the global optimal fitness value of the population, and save the current position of the particle as the global extremum of the population.

Step 8: Speed update

According to the iterative generation P_{ij}, P_{gj} and speed update in Step 7, the speed update formula with additional terms is adopted in this paper, and its formula is shown in (6.19), where $r_3(t)$ is a random number between [0,1].

$$v_{ij}(t+1) = w \cdot v_{ij}(t) + c_1 r_1(t)\big(P_{ij}(t) - x_{ij}(t)\big) \\ + c_2 r_2(t)\big(P_{gj}(t) - x_{ij}(t)\big) + r_3(t)\big(P_{gj}(t) - P_{ij}(t)\big) \quad (6.19)$$

Step 9: Update the solution

Update the solution by the speed generated by Step 8 iteration, that is, adjust the weights and thresholds of BP neural network.
Step 10: Iterative stop control

Evaluate the population generated by iteration, and judge whether the training error of the algorithm reaches the single error (0.001) or the maximum number of iterations (200 generations). If the conditions are met, go to Step 11; Otherwise, return to Step 7 to continue iteration.
Step 11: Optimal solution generation

When the algorithm stops iteration, the corresponding value P_{gj} is the optimal solution of the training problem, that is, the weights and thresholds of BP network. Substitute the optimal solution into BP network model for secondary training and learning, and finally form DDR $(\alpha_1, \beta_1, \alpha_2, \beta_2, \gamma, \sigma)$ prediction optimization model.

6.2.3 *Process of Optimization*

The specific method to optimize the algorithm parameters is shown in Fig. 6.6.

Step 1, Dynamically establishing a simulation model according to historical data of a production line;
Step 2: Establish scheduling rule base, process state (r_h, r_p) and performance index (Move and Utility) required by production line system in simulation model;
Step 3: Determine the bottleneck equipment with equipment utilization rate above 60%;
Step 4: DDR scheduling rules are adopted for bottleneck equipment, corresponding values $(\alpha_1, \beta_1, \alpha_2, \beta_2, \gamma, \sigma)$ are randomly generated respectively, and process state information (r_h, r_p),Move and utility of the production line is automatically recorded at the same time;
Step 5: Apply twice BP neural network algorithm to obtain better value $(\alpha_1, \beta_1, \alpha_2, \beta_2, \gamma, \sigma)$, $\big(r_h, r_p\big)$;
Step 6: Obtain the logical relationship between $(\alpha_1,\ \beta_1,\ \alpha_2,\ \beta_2,\ \gamma,\ \sigma)$ and $\big(r_h,\ r_p\big)$ by linear programming method;

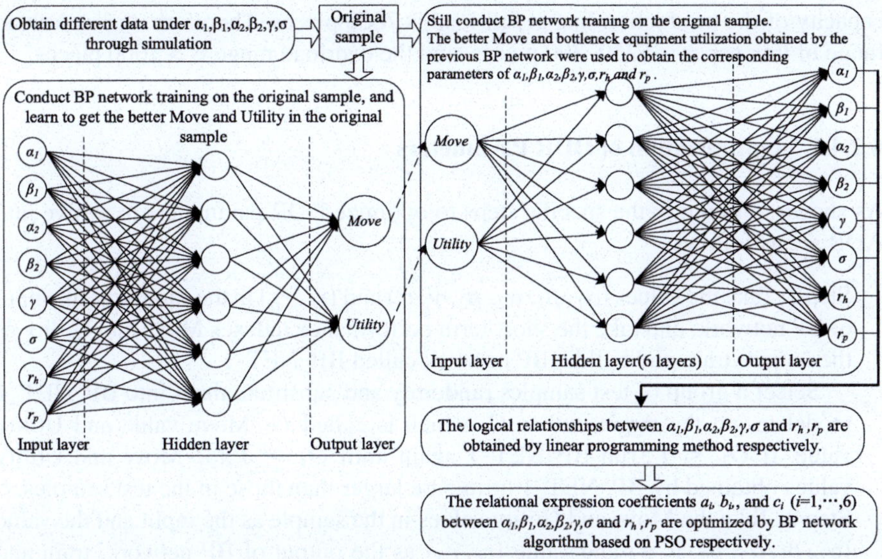

Fig. 6.6 Algorithm parameter optimization based on data mining

Step 7: Using particle swarm optimization neural network algorithm to optimize the coefficient of binary linear relation expression between $(\alpha_1, \beta_1, \alpha_2, \beta_2, \gamma, \sigma)$ and (r_h, r_p).

6.2.4 Simulation and Verification

Based on the historical data of the production line, a certain simulation is carried out, and the equipment with an average utilization rate of more than 60% obtained by simulation for 5 days under the PRIOR rule of the original production line is defined as bottleneck equipment, and DDR is called; Other equipment is still dispatched according to the original scheduling rules.

In this chapter, six parameters $(\alpha_1, \beta_1, \alpha_2, \beta_2, \gamma, \sigma)$ in DDR are cooperatively traversed, with values ranging from 0.01 to 0.99, step size of 0.1, and simulation for 5 days. Among them, the current production line status information (r_h, r_p) is recorded every 12 h, and 110 sets of different values $(\alpha_1, \beta_1, \alpha_2, \beta_2, \gamma, \sigma)$ are simulated to obtain 1100 sets of data.

Among them, 300 sets of data are used for sample verification, 800 sets of data are used for sample training, and cross-validation is adopted to improve the training accuracy.

At the same time, considering that the load state of the production line has a certain influence on the value $(\alpha_1, \beta_1, \alpha_2, \beta_2, \gamma, \sigma)$, the optimization is carried out from two load states: underload and overload. It is known that the daily production capacity of the production line at full load is 7000 pieces. Therefore, the underload range of this paper is 5250–300 pieces, and the overload range is > 7000 pieces.

6.2.4.1 Optimization of DDR Parameters

As shown in Fig. 6.6, the specific steps to optimize DDR parameters to evolve into ADR are as follows:

Step 1: Take the values $(\alpha_1, \beta_1, \alpha_2, \beta_2, \gamma, \sigma)$ and (r_h, r_p) in the sample as the input of BP network, and take the short-term performance indexes Move and Utility as the output, train and build a BP network called BP_NET_1.

Select a group of test samples randomly and substitute them into BP_NET_1 model to get the target output O, which is called O_1 Move value and Utility value. If $O_1 \leq T$, run BP_NET_1 again until $O_1 \geq T$,the Move and Utility values obtained by BP_NET_1 fitting are larger than those in the test samples.

Step 2: Take the Move and Utility values in the sample as the input and the value $(\alpha_1, \beta_1, \alpha_2, \beta_2, \gamma, \sigma)$ and value (r_h, r_p) as the output of BP network, train and build a BP network called BP_NET_2.

At this time, substitute the target output value O_1 obtained in Step 1 into BP_NET_2 model to obtain the target output O, which is called O_2,new value $(\alpha_1, \beta_1, \alpha_2, \beta_2, \gamma, \sigma)$ and value (r_h, r_p).

Finally, the values.$(\alpha_1, \beta_1, \alpha_2, \beta_2, \gamma, \sigma)$ obtained in Step 2 are substituted into the simulation model for verification.

Randomly select three groups of values $(\alpha_1, \beta_1, \alpha_2, \beta_2, \gamma, \sigma)$ when DDR is used for equipment with utilization rate above 60%. After secondary BP network optimization, it is found that (as shown in Tables 6.2 and 6.3), under the underload state, the utility value of Move optimized by secondary BP network is higher than that of Move under the original data, and the utility value is increased by 8.08 and 5.57% on average. At the same time, under the overload condition, the Move, Utility values obtained by the secondary BP network optimization are also higher than those of the original data by 31.07 and 5.57% on average.

Table 6.2 Move value and utility value of secondary BP network optimization (under load)

	α_1	β_1	α_2	β_2	γ	σ	r_h	r_p	Mov	Utility	WIP
Original sample	0.85	0.15	0.7	0.1	0.1	0.1	0.2414	0.1862	10,889	0.6543	5671
	0.05	0.95	0.01	0.01	0.97	0.01	0.3035	0.1900	6552	0.5737	5994
	0.35	0.65	0.4	0.2	0.2	0.2	0.3111	0.1714	12,224	0.7577	6323
Secondary BP	0.49	0.51	0.22	0.36	0.15	0.27	0.2713	0.2168	12,118	0.6891	5671
	0.47	0.53	0.25	0.27	0.26	0.22	0.2955	0.1985	7188	0.6378	5994
	0.42	0.58	0.17	0.44	0.13	0.26	0.3097	0.1632	12,622	0.7677	6323

Table 6.3 Move value and utility value (overload) of secondary BP network optimization

	α_1	β_1	α_2	β_2	γ	σ	r_h	r_p	Mov	Utility	WIP
Original sample	0.5	0.5	0.1	0.3	0.3	0.3	0.2162	0.0990	6202	0.6942	7039
	0.99	0.01	0.3	0.3	0.3	0.1	0.2586	0.0819	4125	0.5613	7395
	0.25	0.75	0.2	0.2	0.4	0.2	0.2539	0.1273	11,591	0.7254	7748
Secondary BP	0.58	0.42	0.07	0.48	0.1	0.35	0.1976	0.1057	8006	0.7054	7039
	0.6	0.4	0.11	0.59	0.11	0.19	0.2418	0.2224	6709	0.6135	7395
	0.55	0.45	0.2	0.45	0.03	0.32	0.2597	0.1305	11,764	0.7674	7748

6.2.4.2 ADR Only Considering Production Line Factors

Based on the BP_NET_2 model, the steps to optimize ADR parameters $(\alpha_1, \beta_1, \alpha_2, \beta_2, \gamma, \sigma)$ and (r_h, r_p) are as follows:

Step 3: Using the method of Linear Programming, LP), the values $(\alpha_1, \beta_1, \alpha_2, \beta_2, \gamma, \sigma)$ and (r_h, r_p) are related with the Formula (6.11).

Because of the problem of large-scale data processing, BP network has good adaptability, self-organization, and fault tolerance. Therefore, in order to better optimize the Move, Utility values, this chapter randomly selects 3–4 groups of samples from 110 groups of data in each 12-h period, optimizes them with the method in 6.2.4.1 to obtain the corresponding value $(\alpha_1, \beta_1, \alpha_2, \beta_2, \gamma, \sigma)$ and (r_h, r_p), and then selects 2–3 groups of $(\alpha_1, \beta_1, \alpha_2, \beta_2, \gamma, \sigma)$ and (r_h, r_p) corresponding to better Move and Utility values from 110 groups of original samples. In this way, four groups of $(\alpha_1, \beta_1, \alpha_2, \beta_2, \gamma, \sigma)$ and (r_h, r_p) are randomly selected from these 5–7 groups of samples at a time, and they are connected by LP method to obtain three groups $(a_i, b_i, c_i, i \in \{1, \ldots, 6\})$, which are substituted into the model and simulated to obtain their corresponding Move, Utility values.

Step 4: Take the Move, Utility values in the 30 groups of sample data obtained in Step 3 as the input of BP network, and $(a_i, b_i, c_i, i \in \{1, \ldots, 6\})$ in the sample data as the output, get the initial weights and thresholds, substitute them into PSO algorithm, optimize the weights and thresholds of BP network by PSO algorithm, and then train them to build a BP network called BP_NET_3.

Same as Step 1, get the target output O, which is called O_3.

Step 5: Take the Move, Utility values in the 30 groups of sample data obtained in Step 3 as the input of BP network and $(a_i, b_i, c_i, i \in \{1, \ldots, 6\})$ in the sample data as the output to obtain the initial weights and thresholds, and use PSO algorithm to optimize the training of BP weights and thresholds to build a BP network called BP_NET_4.

At this time, substitute the target output value O_3 obtained in Step 4 into BP_NET_4 model to obtain the target output O, which is called O_4, namely new $(a_i, b_i, c_i, i \in \{1, \ldots, 6\})$.

Finally, substitute the results $(a_i, b_i, c_i, i \in \{1, \ldots, 6\})$ obtained in Step 5 into the simulation model for verification.

It can be found from Fig. 6.7 that in heuristic scheduling rules, the shortest remaining processing time (srpt) rule has better performance than the Earliest Due Date, EDD) rule and the Critical Ratio, CR) rule in both under-load and overload conditions. However, under the overload condition, the optimized adaptive scheduling rule (ADR) obviously performs better than SRPT rule and is far superior to the PRIOR rule adopted by the enterprise.

Under the underload condition, the improvement effect of DDR is not ideal, and its average value is only 0.64% higher than that of PRIOR rule and 1.03% higher than that of PRIOR rule. The worst DDR result is lower than that of PRIOR rule in both Move and Utility values, while the best DDR result is 4.97% higher than that of prior rule and 7.56% higher than that of prior rule. However, ADR can fully guarantee that its move and utility values are greater than the move and utility values under the PRIOR rule. Among them, the average Move value and the average Utility value of ADR coefficient ($a_i, b_i, c_i, i \in \{1, \ldots, 6\}$) optimized by BP network are increased by 1.97% and 3.26% respectively compared with the Move value and Utility value under PRIOR rule. Using PSO algorithm to optimize the ADR coefficient ($a_i, b_i, c_i, i \in \{1, \ldots, 6\}$) of BP network, its average Move value

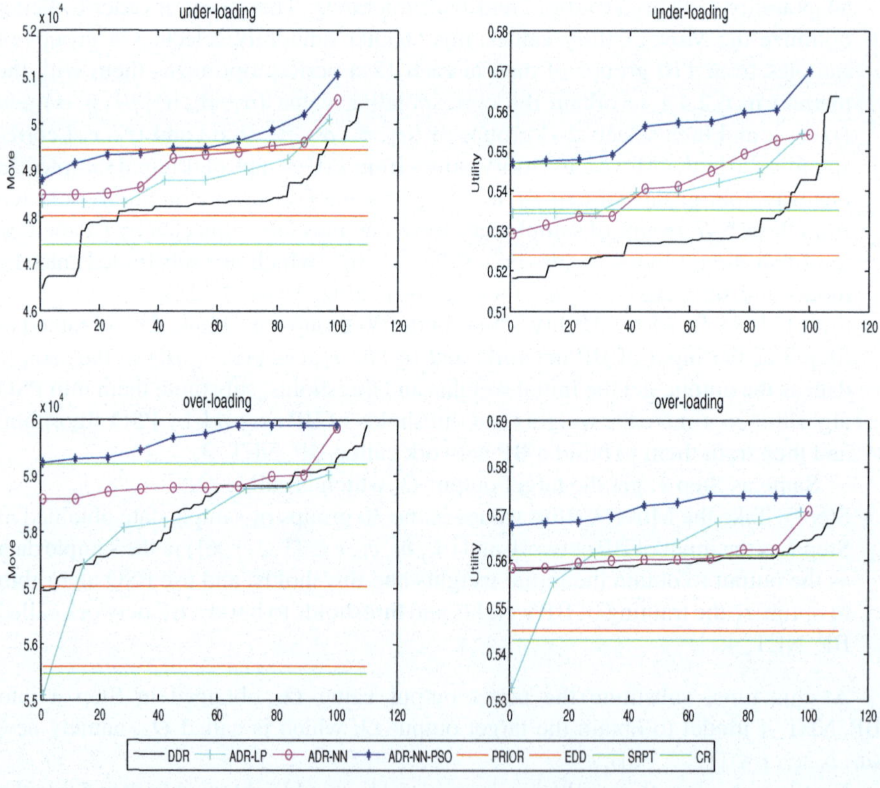

Fig. 6.7 ADR optimization results under under/overload condition

and average Utility value are further improved by 2.35 and 5.93% compared with those under PRIOR rule.

Under overload condition, the Move and Utility values under DDR are better than those under PRIOR rule, and the average move and utility values are increased by 2.87% and 2.11% respectively. Average Move value and average Utility value of ADR coefficient $(a_i, b_i, c_i, i \in \{1, \ldots, 6\})$ optimized by BP network are increased by 5.91% and 2.3 0% respectively compared with Move value and Utility value under PRIOR rule. Using PSO algorithm to optimize the ADR coefficient $(a_i, b_i, c_i, i \in \{1, \ldots, 6\})$ of BP network, its average Move value and average Utility value are further improved by 7.24 and 4.10% compared with those under PRIOR rule.

From the above analysis, it is very necessary to make the value $\alpha_1, \beta_1, \alpha_2, \beta_2, \gamma, \sigma$ in DDR automatically adjust dynamically according to the production line environment and improve it into ADR.

6.2.4.3 ADR Considering Production Line and Lithography Area

Because $\left(r_h, r_p\right)$ are the status information of the production line, they are more common, and in the actual production scheduling process, the bottleneck equipment of non-batch processing is all in the lithography area, so $\left(r_h, r_p\right)$ in the lithography area is stripped from the production line and separately recorded as the emergency workpiece ratio of the lithography area (r_h_photo) and the 1/3 lithography workpiece ratio after the lithography area (r_p_photo).

DDR parameters are optimized in the same way as Step 1 and Step 2, and then the optimized parameter values of single card processing information $(\alpha_1, \beta_1, \alpha_2, \beta_2, \gamma, \sigma)$., $\left(r_h\_photo, r_p\_photo\right)$ and batch processing information $(\alpha_1, \beta_1, \alpha_2, \beta_2, \gamma, \sigma)$, $\left(r_h, r_p\right)$ are fitted by LP respectively, and the Formula (6.20) is obtained.

$$\begin{aligned}
\alpha_1 &= a_1 \cdot r_h_photo + b_1 \cdot r_p_photo + c_1 \\
\beta_1 &= a_2 \cdot r_h_photo + b_2 \cdot r_p_photo + c_2 \\
\alpha_2 &= a_3 \cdot r_h + b_3 \cdot r_p + c_3 \\
\beta_2 &= a_4 \cdot r_h + b_4 \cdot r_p + c_4 \\
\gamma &= a_5 \cdot r_h + b_5 \cdot r_p + c_5 \\
\sigma &= a_6 \cdot r_h + b_6 \cdot r_p + c_6
\end{aligned} \tag{6.20}$$

Then, the new obtained in Step 4 and Step 5 is put into simulation model for verification, and the optimization results are shown in Fig. 6.8, Tables 6.4 and 6.5.

Experiments show that (as shown in Tables 6.4 and 6.5), under the underload condition, considering the lithography area alone, the ADR coefficient $(a_i, b_i, c_i, i \in \{1, \ldots, 6\})$ is optimized by BP network, and its average Move value and average Utility value are increased by 2.39% and 2.11% respectively compared with the average Move value and average Utility value obtained under the overall

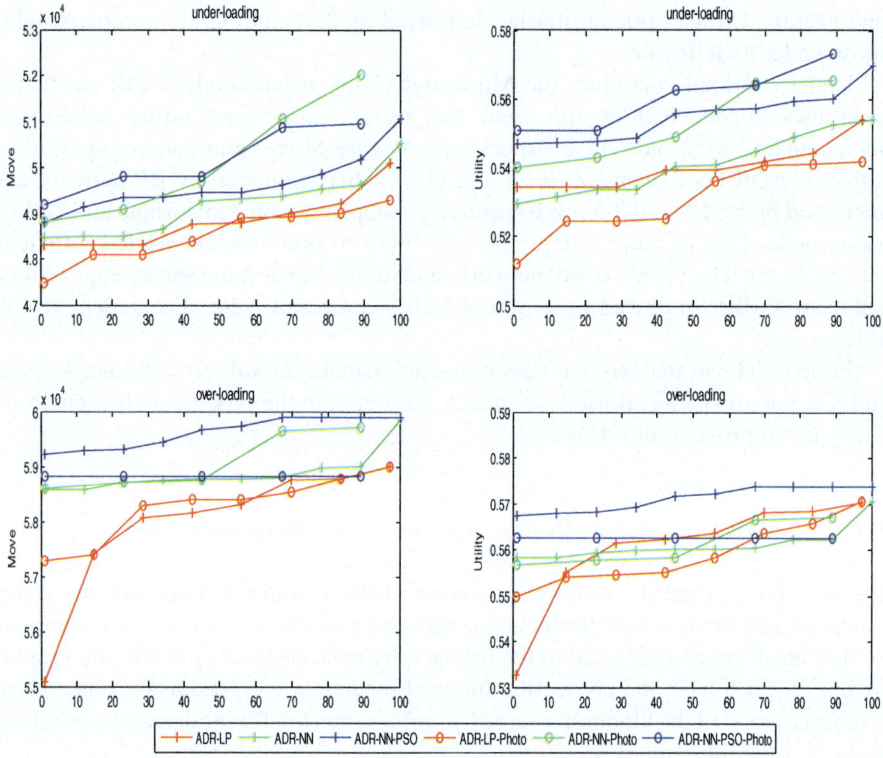

Fig. 6.8 ADR Optimization results of lithography area under under/overload condition

Table 6.4 ADR optimization results under underload condition

Rule	Under-loading				
	Move	Utility	M-Imp. (%)	U-Imp. (%)	A-Imp. (%)
PRIOR	48,027	0.5241	–	–	–
EDD	49,451	0.5388	2.96	2.80	2.88
SRPT	49,639	0.5470	3.36	4.37	3.86
CR	47,419	0.5354	− 1.27	2.16	0.45
DDR best	50,414	0.5637	4.97	7.56	6.26
DDR worst	46,469	0.5182	− 3.24	− 1.13	− 2.18
DDR avg	48,335	0.5295	0.64	1.03	0.84
ADR-LP	48,853	0.5405	1.72	3.13	2.42
ADR-NN	48,975	0.5412	1.97	3.26	2.62
ADR-NN-PSO	49,154	0.5552	2.35	5.93	4.14
ADR-LP-photo	49,083	0.5310	2.2 0	1.32	1.76
ADR-NN-photo	50,146	0.5526	4.41	5.44	4.93
ADR-NN-PSO-photo	50,128	0.5605	4.37	6.95	5.66

Table 6.5 ADR optimization results under overload condition

Rule	Over-loading				
	Move	Utility	M-Imp. (%)	U-Imp. (%)	A-Imp. (%)
PRIOR	55,607	0.5487	–	–	–
EDD	57,043	0.5450	2.58	– 0.67	0.95
SRPT	59,215	0.5614	6.49	2.31	4.40
CR	55,481	0.5429	– 0.23	– 1.06	– 0.64
DDR best	59,900	0.5716	7.72	4.17	5.95
DDR worst	56,942	0.5578	2.40	1.66	2.03
DDR avg	57,205	0.5603	2.87	2.11	2.49
ADR-LP	57,956	0.5604	4.22	2.13	3.18
ADR-NN	58,894	0.5613	5.91	2.30	4.10
ADR-NN-PSO	59,633	0.5712	7.24	4.10	5.67
ADR-LP-photo	58,270	0.5591	4.79	1.9 0	3.35
ADR-NN-photo	59,096	0.5655	6.27	3.06	4.67
ADR-NN-PSO-photo	58,828	0.5627	5.79	2.55	4.17

consideration of the production line. Accordingly, the average Move value and the average Utility value obtained by optimizing BP network with PSO also increased by 1.98% and 0.95%, respectively.

Under the heavy load condition, considering the lithography area alone and optimizing the ADR coefficient $(a_i, b_i, c_i, i \in \{1, \ldots, 6\})$ by BP network, its average Move value and average Utility value are only increased by 0.34% and 0.75% respectively compared with the average Move value and average Utility value obtained under the overall consideration of the production line. However, the average Move value and the average Utility value obtained by optimizing BP network with PSO decreased by 13.8% and 1.51%, respectively, and the effect was not satisfactory. Because there are too many workpieces in the production line under overload condition, the bottleneck equipment always runs at full load, and there is not much room for scheduling optimization. If the production line is always in an overload state, some abnormal situations will easily occur, and no timely response can be obtained, thus reducing the optimization effect.

Generally speaking, ADR associated with status information of lithography area $\left(r_h\_photo, r_p\_photo\right)$ is further adopted for bottleneck equipment in lithography area, while ADR associated with status information of production line $\left(r_h, r_p\right)$ continues to be maintained for other bottleneck equipment, and its Move value and Utility value can be further improved.

6.3 Summary

Based on the actual project of a semiconductor manufacturing enterprise in Shanghai, this chapter proposes a dynamic dispatching rule that simulates pheromone mechanism. This rule breaks through the problem that the traditional intelligent algorithm is limited by the solution scale, and successfully applies it to large-scale complex scheduling problems with good results. At the same time, the data mining method is used to optimize the parameters of dynamic dispatching rules, which are tested and verified in the actual production line.

References

1. Simon H (2004) Neural network principle. Beijing, Mechanical Industry Press
2. Li S, Tiejun C (2009) Intelligent control theory and application. Tsinghua University Press, Beijing
3. Kennedy J, Eberhart RC (1995) Particle swarm optimization. Proceedings of IEEE International Conference on Neural Networks 4:1942–1948

Chapter 7
Performance-Driving Dynamic Scheduling of Semiconductor Manufacturing System

To solve the scheduling problem of semiconductor production line in uncertain production environment, this chapter introduces a performance index driven Dynamic Dispatching Rule (DDR) based on Extreme Leaning Machine, ELM). By means of data mining, the method firstly uses the simulation system of semiconductor production line to simulate, obtains the production data in the production line processing process through simulation, then selects the good parameter group according to the concerned performance index to obtain the required sample set, and then establishes the performance index prediction model of semiconductor production line through the extreme learning machine. Then, through the predicted performance index, combined with the real-time state information of the production line, it learns the best parameters needed by DDR algorithm in the scheduling process, and drives the production line dispatching decision, so that its performance index tends to the predicted value

7.1 Performance Prediction Method

7.1.1 Prediction Method of Long-Term Performances for Single-Bottleneck Semiconductor Production Line

7.1.1.1 Single-Bottleneck Semiconductor Production Line

T Little's Law of classical queuing theory analysis gives a simple mathematical formula about the relationship between WIP quantity and lead time [1–3].

$$L = \lambda W \tag{7.1}$$

L. Li et al., *Data-Driven Scheduling of Semiconductor Manufacturing Systems*, Advanced and Intelligent Manufacturing in China,
https://doi.org/10.1007/978-981-19-7588-2_7

Among them, L refers to the average passenger flow in a stable system for a long time, λ refers to the effective arrival rate of customers in this statistical period, and W refers to the average residence time of each customer in the system.

Little's law requires that the applicable system is stable. Stable means stable production rhythm, products are always output at bottleneck speed, and the speed at which raw materials are put into production line is consistent with bottleneck speed [4]. For a stable system, if the bottleneck speed is constant, the average time spent by each product in the system will be the same. This means that every product can flow out of the system after the same time delay, thus keeping the output speed and arrival speed of the system synchronized [5]. For semiconductor manufacturing systems, this requires stable bottleneck links in the whole production process, fixed bottleneck equipment and stable output, or all bottleneck links in the production line can be classified as a single bottleneck [6]. In this paper, the production situation conforming to this characteristic is summarized as a single bottleneck semiconductor production model, at which time the number of WIP on the production line can generally maintain a stable level [7].

According to the obtained data of actual semiconductor production line, it can be observed from the comparison chart of daily WIP quantity and daily queue length in Fig. 4.4 of Chap. 4 that the actual WIP level of the production line is basically stable for most of the year. In this section, we assume the production situation of the production line, that is, it is assumed that the production line is a single bottleneck semiconductor production model with constant production tempo. Each bottleneck link in the production process can be simplified as a single bottleneck, the output speed is uniform, the releasing speed keeps pace with it, and the production tempo keeps the same as the production speed of the bottleneck equipment after being classified [8].

Based on the hypothesis of single bottleneck semiconductor production model, the quantitative linear relationship between processing cycle (CT)/on-time delivery rate (ODR) and key short-term performance indicators will be established below, so that the short-term performance indicators of current working conditions can be reflected by real-time collected data, and the key long-term performance indicators of a certain product-processing cycle and on-time delivery rate can be predicted [9].

7.1.1.2 Multiple Linear Regression Problem

In statistics, Linear Regression is one of regression analysis methods, which refers to modeling the relationship between one or more independent variables and dependent variables by using linear regression equations [10]. Linear regression function is a linear combination of one or more model parameters called regression coefficients. The variables x, y in Fig. 7.1 are considered to be approximately linear.

Univariate linear regression analysis means that the model includes only one independent variable and one dependent variable, and the relationship between them can be expressed by a straight line approximately. Multiple linear regression analysis

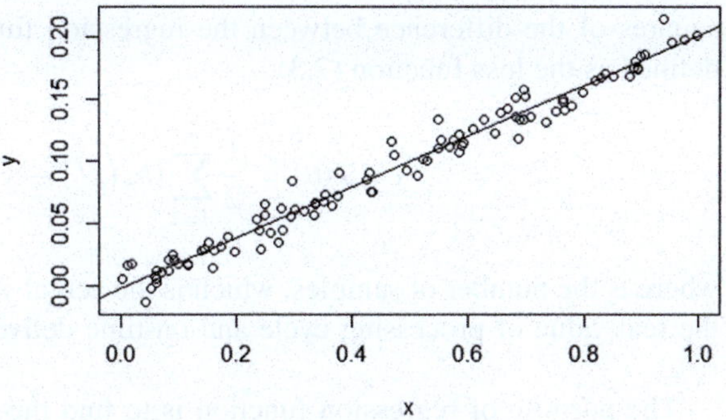

Fig. 7.1 Schematic diagram of linear relationship

means that the model includes two or more independent variables, and the relationship between dependent variables and independent variables is approximately linear [11].

In regression analysis, the choice of independent variables plays an important role in ensuring that the multiple linear regression model has good prediction effect and excellent interpretation ability. Independent variables should have a close linear correlation with dependent variables and have a strong logical influence, and independent variables should have complete statistical data.

In the single bottleneck semiconductor production model, the system conforms to Little's law. Under such production conditions, there is a linear relationship between long-term performance index and short-term performance index variables, and the prediction model is often realized based on linear regression in engineering. In Chap. 4, the correlation analysis verifies the correlation between each short-term performance index and the daily moving steps MOVE, which directly reflects the production line efficiency, and basically meets the above criteria. Therefore, the long-term performance index prediction problem of single bottleneck semiconductor production model is reduced to a multiple linear regression problem.

The commonly used methods to solve multiple linear regression problems are least square method and gradient descent method. The least square method is an idea of optimization problem, and the gradient descent method is a concrete solution method to realize this optimization idea. In this paper, gradient descent method is chosen to solve the multiple linear regression problem [12].

Assume that the regression function is:

$$h(x) = \sum_{i=1}^{n} \theta_i x_i = \theta^T x \tag{7.2}$$

where is the number of independent variables and the coefficient of independent variables, that is, the weight of each short-term performance index that needs to be solved as the input of the model. The mean square of the difference between the regression function. The actual value is defined as the loss function: nθ_i is the coefficient of the independent variable, that is, the weight of each short-term performance index that needs to be solved as the model input. The sum of the mean squares of the difference between the regression function and the actual value is defined as the loss function (7.3).

$$J(\theta) = \frac{1}{2}\sum_{j=1}^{m}\left(h_\theta(x^j) - y^j\right)^2 \tag{7.3}$$

where is the number of samples, which is the actual value in the training set, that is, the real value of processing cycle and on-time delivery rate. The true value of the rate.

The purpose of regression function is to find the value of the parameter θ that minimizes the loss function $J(\theta)$. For each parameter θ_i, find its gradient expression and make the gradient equal to 0 to find θ_i. At this time, the obtained parameter θ_i minimizes the loss function.

θ is a one-dimensional vector containing all parameters. Initialize one θ first, above this value θ, calculate the value of the next group θ by using the random gradient descent method, and with the update of θ, the value of the loss function $J(\theta)$ decreases continuously. When iterated to a certain extent, the value of $J(\theta)$ tends to be stable, and then θ_i is the required value. The iterative function is as follows, where α is the step size of gradient descent:

$$\theta_i = \theta_i - \alpha\frac{\partial}{\partial\theta}J(\theta) \tag{7.4}$$

For each iteration, use the current θ_i to find the value on the right side of the equation, and overwrite the original value θ_i after iteration.

$$\frac{\partial}{\partial\theta_i}J(\theta) = (h_\theta(x) - y)x_i \tag{7.5}$$

In the above solution, the relationship between long-term and short-term performance indexes is established only by training historical data. As a matter of fact, it can be considered that the error E obeys Gaussian distribution. By calculating the expectation and variance of the error, the Gaussian distribution function of the error is obtained. The error function is used to fit the future error.

$$e \sim N\left(\mu, \sigma^2\right) \tag{7.6}$$

$$\begin{aligned}\hat{\mu} &= \tfrac{1}{n}\sum_{i=1}^{n} x_i \\ \hat{\sigma}^2 &= \tfrac{1}{n}\sum_{i=1}^{n}(x_i - \overline{x})^2\end{aligned} \tag{7.7}$$

The predicted value of the final long-term performance index is the predicted value $h(x)$ obtained by multiple linear regression model plus error compensation e fitted by Gaussian distribution function.

$$y_i = h(x) + e \tag{7.8}$$

7.1.1.3 Prediction Model of Processing Cycle

(1) Basic prediction model of processing cycle

Based on the above discussion on the long-term performance index prediction of single bottleneck semiconductor production model, this section regards the processing cycle (CT) as a dependent variable and the corresponding short-term performance index WIP_t, QL_t, MOVE_t, TH_t, EQI_UTI_{1-18} as an independent variable. The gradient descent method described in the previous section is used to train the model, and the relational equation of the processing cycle prediction model is obtained.

Table 7.1 is the basic prediction model of processing cycle. The parameters represent the coefficients multiplied by each short-term performance index in the table, and these short-term performance indexes are multiplied by the corresponding coefficients and summed to obtain the predicted value of the processing cycle of this version of products under the current working conditions.

(2) Error parameters of machining cycle prediction model

The prediction model of machining cycle with error compensation is obtained by adding fitting error compensation prediction to the basic relationship model obtained by multiple linear regression method. Among them, the model coefficients of machining cycle relationship are the same as the original, but the errors are regarded as conforming to Gaussian distribution. During training, Gaussian distribution parameters for fitting future errors are calculated at the same time. At this time, the predicted value of the last machining cycle is the sum of the predicted value of the basic model and the predicted error compensation. Gaussian distribution parameters of processing cycle error of each version of products are shown in Table 7.2, μ is the expected value of error distribution and σ^2 is the variance.

Table 7.3 shows the test results of the processing cycle relationship model. The original error rate indicates the prediction results of the model when Gaussian distribution is not used to fit future errors, and the error rate after error compensation is the prediction result obtained by adding error function. Where, error rate = | actual value-predicted value |/actual value.

Table 7.1 P4 processing cycle relationship model

Serial number	Parameter	Short term performance index
k0	0.00703	WIP
k1	− 0.022	MOVE
k2	0.031246	QL
k3	− 0.00761	Throughput
k4	− 0.08883	2CL01
k5	− 0.0631	7MF04
k6	− 0.0178	9CL20
k7	− 0.14287	5853
k8	− 0.11527	1703
k9	− 0.02277	5854
k10	− 0.00845	9PS18
k11	− 0.04355	5856
k12	− 0.05178	5852
k13	0.034341	2CL05
k14	0.159431	6DI02
k15	0.039893	3WE10
k16	0.002016	3T05
k17	− 0.24397	6113
k18	− 0.06953	5821
k19	− 0.0111	2T03
k20	− 0.13699	6DI01
k21	− 0.09838	6148

Table 7.2 Gaussian distribution parameters of error compensation in processing cycle of each product

Product version	μ	σ^2
P1	0.742001	397.7432
P2	− 0.18178	396.6348
P3	0.447137	137.2231
P4	0.597158	7080.552
P5	− 0.00336	4.745342
P6	2.990688	923.4504
P7	1.338826	317.1728
P8	0.727428	40.52799

Table 7.3 Test results of processing cycle prediction model

Product version number	Original error rate (%)	Error rate after compensation (%)	Improvement rate (%)
P1	36.80	28.03	– 10.70
P2	19.10	13.50	– 16.98
P3	28.60	26.38	– 6.20
P4	31.80	30.30	17.66
P5	26.21	17.57	8.01
P6	25.32	25.26	11.68
P7	11.54	25.71	19.15
P8	24.84	24.87	5.11

Figure 7.2 is a column comparison diagram of error rate between prediction without error and prediction with error. According to the test results, it can be seen that after adding error compensation, the prediction accuracy of processing cycle of five kinds of products has improved, but the prediction accuracy of three kinds of products has decreased, that is, the accuracy of nearly half of products has decreased. Therefore, on the whole, considering the error as Gaussian distribution is not very reasonable for predicting the product processing cycle.

Therefore, this section uses the basic model without error prediction as the final prediction result. Table 7.4 shows the comparison table between the predicted value and the true value of the product processing cycle of each version. Due to the space limitation, three values are randomly selected for each version. Figure 7.3 is the radar chart of deviation between the predicted value and the true value predicted by the

Fig. 7.2 Error rate comparison between basic relationship model of machining cycle and relationship model with error prediction

Table 7.4 Comparison table of actual value and predicted value of product processing cycle in different versions

Product version number	Predicted value	Actual value
P1	13.99324	14
	12.23443	13
	12.8143	10
P2	12.07208	11
	15.13227	22
	8.344726	12
P3	9.763301	10
	15.34756	18
	10.96271	17
P4	12.952	10
	12.86826	21
	8.503999	11
P5	19.94438	21
	4.764723	4
	4.768187	6
P6	26.98716	31
	25.61281	18
	15.55442	20
P7	21.05982	26
	21.05982	17
	21.05982	17
P8	21.05982	26
	15.37907	15
	26.26116	23

basic relationship model of product processing cycle of each version of the test set. Blue lines represent predicted values and orange lines represent actual values.

7.1.1.4 Prediction Model of On-Time Delivery Rate

(1) Forecasting model of on-time delivery rate

Taking on-time delivery rate (ODR) as dependent variable and corresponding short-term performance indexes WIP_t, QL_t, MOVE_t, TH_t as independent variable, the model is solved by gradient descent method, and the relational equation of on-time delivery rate prediction model is obtained.

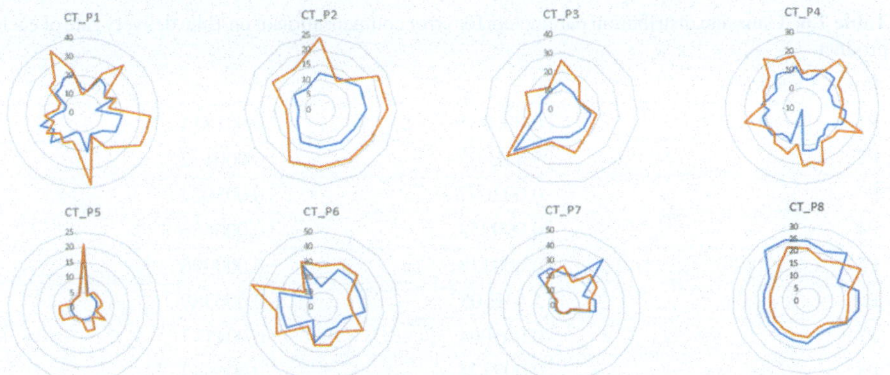

Fig. 7.3 Radar chart of deviation between predicted value and actual value CT

Table 7.5 is the basic forecasting model of on-time delivery rate ODR of P4 products. The parameter represents the coefficient multiplied by each short-term performance index. These short-term performance indexes are multiplied by the corresponding coefficients and added together, so that the predicted value of the on-time delivery rate of this version of products under the current working conditions can be obtained.

(2) Error parameters of on-time delivery rate prediction model

Same as above, add fitting error compensation to the basic relation model of on-time delivery rate based on multiple linear regression method, and get the prediction model of on-time delivery rate with error compensation. Here, the errors are regarded as obeying Gaussian distribution, and the Gaussian distribution parameters for fitting future errors are calculated during training, as shown in Table 7.6.

Table 7.7 shows the test results of the on-time delivery rate prediction model. Like the machining cycle, the original error rate represents the prediction result of the basic model when the future error is fitted without Gaussian distribution. The error rate after adding error compensation is the prediction result after adding error function. Where, error rate = | actual value-predicted value |/actual value.

Table 7.5 P4 on-time delivery rate relational model

Serial number	Parameter	Short term performance index
k0	0.00125	WIP
k1	5.42E-05	MOVE
k2	− 0.00163	QL
k3	0.000809	Throughput

Table 7.6 Gaussian distribution parameters for error compensation of on-time delivery rate of each product

Product version	μ	σ^2
P1	0.000935	0.001814
P2	5.45E-05	0.000192
P3	0.002697	0.003027
P4	0.000253	0.000656
P5	0.000118	0.000996
P6	5.43E-05	0.000389
P7	0.001064	0.004117
P8	0.005174	0.006642

Table 7.7 Test results of on-time delivery rate prediction model

Product version number	Original error rate (%)	Error rate after compensation (%)	Improvement rate (%)
P1	11.02	10.96	0.54
P2	2.99	2.79	6.69
P3	5.44	5.01	7.90
P4	27.54	24.56	10.82
P5	24.10	20.51	14.90
P6	10.70	10.69	0.09
P7	15.91	12.96	18.54
P8	7.83	7.02	10.34

Figure 7.4 is a comparison histogram of error rates of the relational models without error prediction and with error prediction. It can be seen from the figure that the error rate of ODR has been significantly improved after adding error compensation. This may be because there are more factors affecting the processing cycle than those affecting the on-time delivery rate, and Gaussian distribution cannot describe the former error well.

Therefore, based on the hypothesis of single bottleneck semiconductor production model, the just-in-time delivery rate relationship model with error prediction based on multiple linear regression method is selected as the final prediction model. Table 7.8 shows the comparison table between the predicted value and the true value of the on-time delivery rate of each version. Due to the space limitation, each version takes three values randomly. Figure 7.5 is a radar chart of the predicted value and the true value of the on-time delivery rate relationship model of each version of products. Blue lines represent predicted values and orange lines represent actual values.

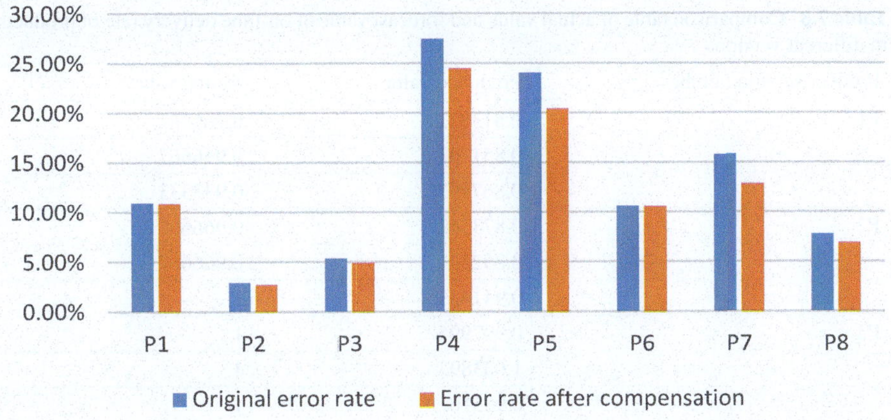

Fig. 7.4 Error rate comparison diagram of basic relation model of just-in-time delivery rate and prediction model with error

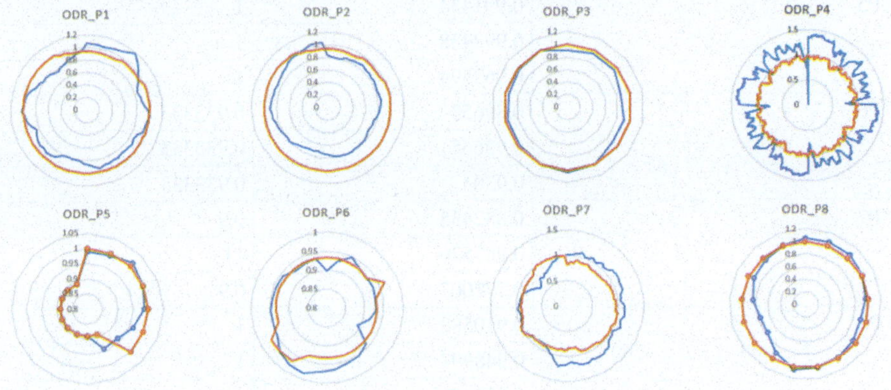

Fig. 7.5 Radar chart of deviation between predicted value and actual value of ODR

7.1.2 Prediction Method of Long-Term Performances for Multi-bottleneck Semiconductor Production Line

7.1.2.1 Multi-bottleneck Semiconductor Production Line

In many real semiconductor manufacturing systems, orders are highly uncertain, and are limited by the number of equipment, product process requirements, and the ratio of equipment in each processing area. Bottleneck equipment exists in many links of the production line. Because the occurrence time of bottleneck is uncertain, the output of each bottleneck device is unstable, which makes it impossible to simplify all bottleneck devices in the whole production line into a bottleneck node model

Table 7.8 Comparison table of actual value and forecast value of on-time delivery rate of products in different versions

Product version number	Predicted value	Actual value
P1	0.813822	0.966667
	0.851731	0.966667
	0.867497	0.933333
P2	0.873362	0.966667
	0.852321	0.966667
	0.941435	1
P3	0.997995	1
	1.033893	1
	0.99516	1
P4	0.992356	0.966667
	0.977561	0.966667
	0.953224	1
P5	0.991434	1
	0.992438	1
	0.963594	1
P6	0.89653	0.933333
	0.92665	0.933333
	0.9303	0.933333
P7	0.882435	0.9
	0.987526	11
	0.772007	0.9
P8	0.910588	1
	0.948997	1
	0.92545	1

in many cases, and the actual releasing cannot keep pace with the uncertain multi-bottleneck production speed. The uncertainty effect of multiple bottleneck nodes will extend to the whole production line, and finally the products will not be produced at uniform bottleneck speed. This kind of production situation is summarized as a multi-bottleneck semiconductor production model.

At this time, the system may no longer be a stable linear system, and its output speed is often inconsistent with its arrival speed, which makes the long-term performance indexes of various products fluctuate greatly. In this case, the long-term performance indicators and short-term performance indicators reflecting the current working conditions may not be well expressed by linear relationship, and their actual relationship may no longer conform to a certain statistical law in a certain plane dimension, but may penetrate multiple dimensions and correlate with each other in a function-space view. The actual production data set used in this paper may also conform to this production model. For this kind of production with unstable bottleneck output, we tend to seek a more complex modeling method of probability distribution in high dimension, and model the inline relationship between long-term and short-term performance indicators on the production line. In this section, based on the multi-bottleneck semiconductor production model, a prediction model based on Gaussian process regression method is proposed to predict and model two long-term performance indicators. The actual data set of a semiconductor in Chap. 2 is also taken as the research sample, and the prediction results are compared with those of the above-mentioned multiple linear regression prediction method based on the hypothesis of single-bottleneck semiconductor production model, so as to explore which production model the semiconductor production line is more suitable for.

7.1.2.2 Gauss Process Regression Problem

A group of random variables with certain correlation and independence is called random process. Random process can be regarded as a collection of many random variables, which represents the change of a random system with an indicator vector. This indicator vector is usually a physical quantity, such as time. Traditional probability theory usually studies the relationship between one or more independent random variables. In the law of large numbers and the central limit theorem, infinite random variables are studied, but they are still based on the assumption that these random variables are independent of each other. In a multi-bottleneck semiconductor production model with multi-product lines and bottlenecks unable to maintain stable and uniform output, the relationship between short-term performance indicators is not independent, so the relationship between long-term performance indicators and short-term performance indicators can be summarized as a random process.

Random process can be defined by a random variable cluster $X(t, w), t \in T$. . Gaussian process regression can be regarded as the expansion of multidimensional Gaussian distribution to infinite dimension, and can be interpreted as the distribution of functions, that is, the function-space relationship. Different from other random processes, in the Gaussian process, a limited number of indexes (such as n, $t_1, t_2, \ldots, t_n$) are randomly selected from random variable clusters, and the joint distribution of vectors t composed of corresponding random variables is multidimensional (such as n-dimensional) Gaussian distribution. Specifically, each point in the input space is associated with a random variable that obeys Gaussian distribution. At the same time, for any combination of finite random variables, the joint probability also obeys Gaussian distribution. When the indicator vector t is two-dimensional or multidimensional, the Gaussian process becomes a Gaussian random field. Gaussian process regression has been widely used in probability and statistics theory. Figure 7.6 is a schematic diagram of the sampling function generated by a Gaussian regression process. From the posterior function, three random functions that conform to Gaussian distribution can be obtained based on the conditional prior probability distribution of five noiseless observations. The shaded part represents the confidence

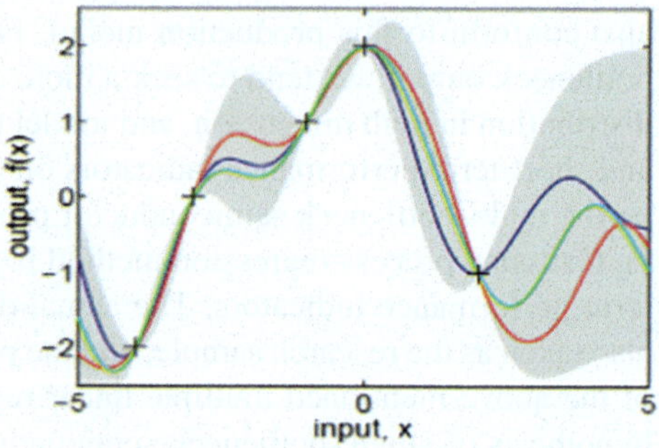

Fig. 7.6 Gaussian regression process

interval of adding and subtracting the mean value by 2 times the standard deviation for each input value (taking 95% as the confidence interval).

Generally, the Gaussian process is expressed by means of mean and covariance [54]. Many research assume that the mean value m is zero when modeling with Gaussian process method $f \sim \mathrm{GP}(m, K)$, and then determine the covariance function K according to the specific application.

In this paper, the mean and covariance of Gaussian process $f(x)$ are defined as:

$$\begin{aligned} m(x) &= \mathbb{E}[f(x)] \\ k\left(x, x'\right) &= \mathbb{E}\left[(f(x) - m(x))\left(f\left(x'\right) - m\left(x'\right)\right)\right] \end{aligned} \tag{7.9}$$

Generally, the function $f(x)$ is assumed to be given a prior Gaussian process, and the Gaussian process can be regressed as follows:

$$f(x) \sim GP\left(m(x), k\left(x, x'\right)\right) \tag{7.10}$$

Same as the existing method, for the convenience of calculation and expression, the mean value is set as zero here. Covariance function can describe the correlation between given points. In other words, this is a function used to measure similarity. Choose the Squared Exponential function here:

$$k\left(x, x'\right) = \exp\left(-\frac{1}{2}\left|x - x'\right|^2\right) \tag{7.11}$$

When the input variables are closely related, the covariance between the variables increases. On the contrary, when the distance between input variables increases, that is, the correlation decreases, the covariance also decreases accordingly. This is the basic hypothesis of Gaussian process regression model in predicting unknown variables. That is, when the working conditions are similar, the corresponding long-term performance index values should also be similar. In the prediction problem, when a given training set $x_1, x_2, \ldots, x_n$ and the corresponding function value $f_1, f_2, \ldots, f_n$:

$$D = \left\{x^{(i)}, f^{(i)} \middle| i = 1, \ldots, n\right\} = \{X, f\} \tag{7.12}$$

The goal of the function is to calculate the value of the corresponding predicted variable f^* when the input test point is x^*. According to the above-mentioned properties of Gaussian process, that is, the test data and training data come from the same distribution, the joint probability distribution of training data f and test data f^*, that is, high-dimensional Gaussian distribution, can be obtained:

$$\begin{bmatrix} f \\ f^* \end{bmatrix} \sim \mathcal{N}\left(0, \begin{bmatrix} k(X,X), k(X,X^*) \\ k(X^*,X), k(X^*,X^*) \end{bmatrix}\right) \tag{7.13}$$

The training set is the observed value. Therefore, the problem of solving prediction data can be simplified to the problem of calculating its posterior probability based on observable values. In probability theory, this calculation can be converted into calculating the conditional joint Gaussian posterior distribution based on observations:

$$\begin{aligned} & f^* D = \{X, f\} \\ & f^*(X^*, X, f \sim N(k\left(k(X^*, X)k(X, X)\right) f, k(X^*, X^*) \\ & \quad - k(X^*, X)k(X, X)k\left(X, X^*\right)) \end{aligned} \tag{7.14}$$

In other words, the posterior function $P(f^*|X^*, \mathrm{X}, f)$ is a Gaussian distribution with the following mean and covariance:

$$\begin{aligned} & \mu = k(X^*, X)k(X, X)^{-1} f \\ & \sum k(X^*, X^*) - k(X^*, X)k(X, X)^{-1} k(X, X^*) \end{aligned} \tag{7.15}$$

For the test set, the predicted value to be solved is the expectation μ of f^*, that is, the mean value of distribution, which can be easily calculated by the above equation.

7.1.2.3 Prediction Model of Processing Cycle

This section is based on the hypothesis of multi-bottleneck semiconductor production model, that is, the bottleneck equipment exists in many links of the production line, which can not be simplified into a single bottleneck, and the output is unstable, and the actual releasing speed and bottleneck output speed cannot be synchronized. Currently, the system is no longer a stable linear system, and each long-term and short-term performance index has a certain correlation, which is not independent and affects each other. In view of this typical stochastic system, this section adopts Gaussian process regression method to predict and model the processing cycle and various short-term performance indexes related to the production line. Each short-term performance index in the input space is regarded as a random variable that obeys a Gaussian distribution, and the joint probability of these short-term performance index combinations also obeys a Gaussian distribution.

From The training set of actual semiconductor production line processing cycles, the processing cycles of completed workpieces and their corresponding short-term performance indexes according to different products are selected as model inputs, including the daily WIP of the whole production line, the daily queue length QL, the daily total moving steps MOVE, the sunrise piece quantity th, and the annual average equipment utilization rate of 18 screened representative equipment. In the above joint probability distribution:

$$\begin{bmatrix} f \\ f^* \end{bmatrix} \sim \mathcal{N}\left(0, \begin{bmatrix} k(X, X^*), k(X, X^*) \\ k(X^*, X), k(X^*, X^*) \end{bmatrix}\right) \tag{7.16}$$

Among them, X represents 22 observed short-term performance indexes, f represents the actual processing cycles of each finished workpiece, and the dimension of X is 22. It can be found that the Gaussian process regression model is not only concise in expression, but also convenient in calculation.

In the previous section, based on the assumption of single bottleneck semiconductor production model, multiple linear regression method was used to predict and model the long-term performance index. In this experiment, the prediction results of multiple linear regression method are compared with those based on Gaussian process regression method. By comparing the different prediction algorithms based on two production model assumptions on the same real production line data, the conclusion is drawn.

Gaussian process regression can directly give the predicted value of processing cycle of each product in the test set. In the previous processing cycle prediction based on multiple linear regression, we compared the processing cycle prediction results without error compensation with those with error compensation, and got the conclusion that when modeling the processing cycle based on multiple linear regression under the assumption of single bottleneck semiconductor production model, some product lines choose the method with error compensation to have higher prediction accuracy, while some product lines choose the model without error compensation to have more accurate prediction. In the comparative experiment here, the best prediction results of each product line based on multiple linear regression method are selected as the benchmark experiment and compared with the experimental results obtained by the prediction method based on Gaussian process regression.

Table 7.9 shows the comparison between the prediction results based on Gaussian process regression model and the best prediction results based on multivariate linear regression model for each product line. Figure 7.7 shows the columnar comparison chart of error rate. It can be seen from the chart that the prediction accuracy of Gaussian process regression prediction model for each product has been greatly improved compared with multiple linear regression. This is because Gaussian process regression makes predictions by comparing the observed values in the test set and training set with their neighbors, and modeling assumes that if the current working condition is similar to the historical working condition, then the machining cycle should be similar. This modeling method can well express the production status of semiconductor manufacturing system in reality and is more suitable for the characteristics of data sets. Figure 7.8 is the radar chart of deviation between the predicted value and the real value predicted by the relationship model between each version of product processing cycle and each short-term performance index. Blue lines represent predicted values and orange lines represent actual values.

Table 7.9 Comparison table of processing cycle prediction results based on two prediction methods

Product version number	Optimal prediction error based on LR method (%)	Prediction error based on GPR method (%)	Improvement rate (%)
P1	25.32	9.55	62.28
P2	11.54	9.59	16.90
P3	24.84	10.01	59.70
P4	30.30	12.33	59.31
P5	17.57	14.04	20.09
P6	25.26	12.98	48.61
P7	25.71	14.79	42.47
P8	24.87	4.73	80.98

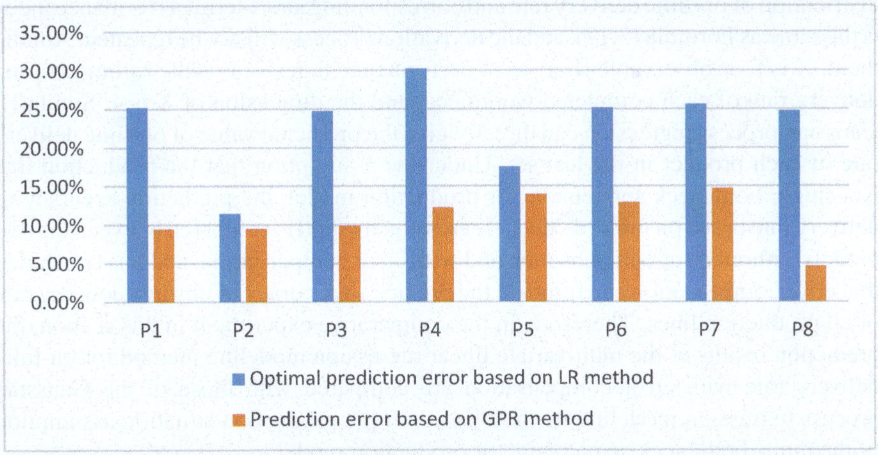

Fig. 7.7 Comparison chart of optimal error rate between Gaussian process regression model and multivariate linear relationship model in machining cycle

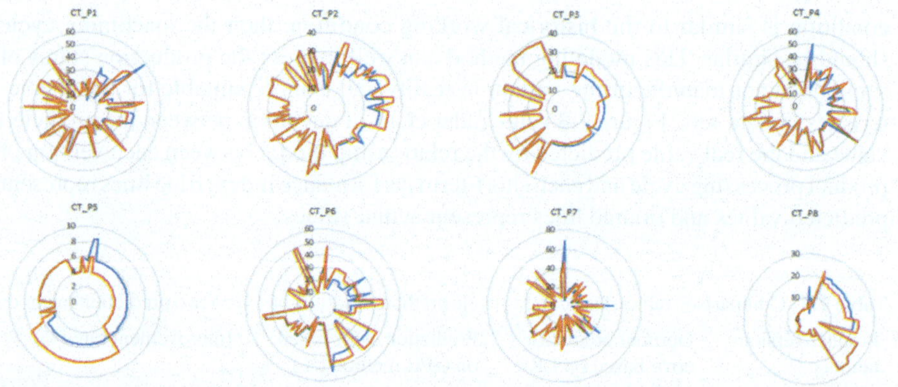

Fig. 7.8 Radar chart of deviation between predicted value and actual value of CT under multiple bottlenecks

7.1.2.4 Prediction Model of On-Time Delivery Rate

Consistent with the algorithm in the previous section, in the on-time delivery rate data set, the on-time delivery rate of the product version of the finished workpiece in its average processing cycle and the four corresponding short-term performance indicators (WIP_t, QL_t, MOVE_t, TH_t) of the production line in the starting date of the statistical cycle have been selected as model inputs. The joint probability distribution of on-time delivery rate and corresponding short-term performance index is the same as Formula (3.17), and the derivation process will not be repeated. Among them, X is four observable short-term performance indexes, F is the rolling on-time delivery rate of each completed workpiece, and the dimension of X is 4. Similarly, Gaussian process regression can directly give the predicted value of on-time delivery rate of each product in the test set. Under the assumption that the production line is a single bottleneck semiconductor production model, the prediction accuracy of delivery rate based on multiple linear regression method is compared between the two models without error compensation and with error compensation, and it is concluded that error compensation can improve the prediction accuracy of on-time delivery rate of all production lines. Therefore, in the comparative experiment in this section, the prediction results of the multivariate linear regression modeling method for on-time delivery rate with error compensation are compared with those of the Gaussian process regression modeling method based on the production situation assumption of the multi-bottleneck semiconductor production model.

Table 7.10 shows the comparison results between the on-time delivery rate prediction based on Gaussian process regression model and multivariate linear regression model with error prediction proposed in this section. Figure 7.9 shows a columnar comparison diagram of the error rates of the two prediction methods. Figure 7.10 is a radar chart of deviation between predicted and true values of on-time delivery rate of each version of products based on Gaussian process regression. Blue lines represent predicted values and orange lines represent actual values.

According to the experimental results, the following conclusions can be drawn:

(1) In the forecasting of on-time delivery rate, the forecasting method based on Gaussian process regression is also more effective than the forecasting model based on multiple linear regression. The experimental results fully show that the actual data set of semiconductor production line is more in line with the hypothesis of multi-bottleneck semiconductor production model, that is, the bottleneck equipment is scattered in all links of the production line, resulting in unstable bottleneck output. This makes the potential association mode among bottleneck devices unstable, and it is impossible to simplify all bottleneck devices of the system into a single bottleneck node model from the data mode level. The actual

Table 7.10 Comparison of on-time delivery rate prediction results based on two prediction methods

Product version number	Prediction error based on LR method (%)	Prediction error based on GPR method (%)	Improvement rate (%)
P1	10.96	3.14	71.35
P2	2.79	1.08	61.29
P3	5.01	2.04	59.28
P4	24.56	4.25	82.70
P5	20.51	4.30	79.03
P6	10.69	4.35	59.31
P7	12.96	2.54	80.40
P8	7.02	2.07	70.51

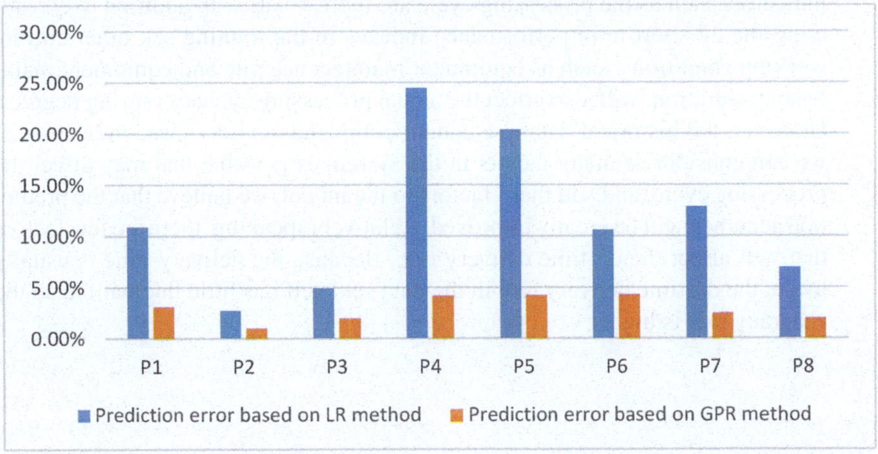

Fig. 7.9 Error rate comparison between Gaussian process regression model of on-time delivery rate and multivariate linear relationship model with error prediction

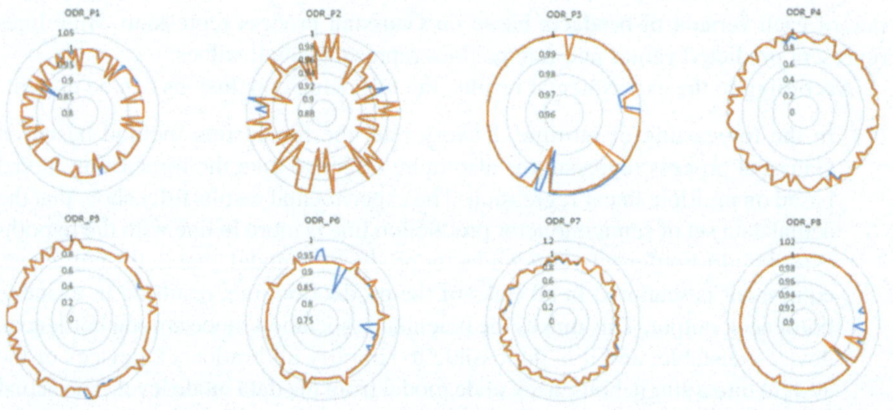

Fig. 7.10 Radar chart of deviation between predicted value and actual value of ODR under multiple bottlenecks

releasing speed cannot keep pace with the uncertain multi-bottleneck production speed. Gaussian process regression method makes prediction by comparing the observation values near the test set and the training set, and models based on the assumption that the predicted processing cycle and on-time delivery rate should be similar if the current working condition is similar to the historical working condition. The experimental results show that the modeling method based on this assumption can well express the production status of the semiconductor manufacturing system in reality and is more suitable for the data set characteristics of this production system.

(2) The accuracy of the on-time delivery rate ODR of the same product is much higher than that of the processing cycle, because the long-term performance indicators such as the processing cycle are more volatile. In addition to considering the 22 short-term performance indexes in the training set, other current working conditions, such as equipment maintenance rate and equipment maintenance duration, will also affect the actual processing cycle to varying degrees. However, the historical database cannot completely extract these indicators. If we can consider as many factors in the system as possible that may affect the processing cycle, and add these factors to the model, we believe that the prediction accuracy will be greatly improved. Relatively speaking, there are few factors that will affect the on-time delivery rate. Because the delivery time is usually loose, the on-time delivery rate in the data set itself has little fluctuation, so the accuracy rate is higher.

7.2 Dynamic Scheduling of Semiconductor Production Line Based on Load Balancing

There are many multiple entry processes in semiconductor production lines, and some equipments are usually very expensive, so the equipment load on the production line is usually heavy. The proposed load balancing model can well adjust the load distribution of the production line and improve the overall performance of the production line [13]. This chapter introduces a closed-loop optimized dynamic dispatching rule considering load balancing (DDR LB).

7.2.1 Overall Design

Figure 7.11 shows the structure diagram of the dynamic scheduling method for closed-loop optimization of semiconductor production line based on load balancing

In this section, the semiconductor production line is taken as the research object, and from the perspective of closed-loop optimal scheduling, the closed-loop optimal scheduling strategy of the semiconductor production line is designed and implemented, to achieve targeted, controllable and rapid dispatching of production line workpieces, and make the whole production system reach a dynamic equilibrium state, thereby improving productivity and operational performance [14]. The scheduling method includes the following three aspects:

(1) The establishment of dynamic balance equation, that is, the establishment of dynamic balance model of semiconductor production line scheduling system, and the functional relationship between performance index (output) and expected performance (reference value);
(2) The study of dynamic control strategy (controller), that is, the establishment of closed-loop control strategy, makes the whole production line achieve dynamic balance;

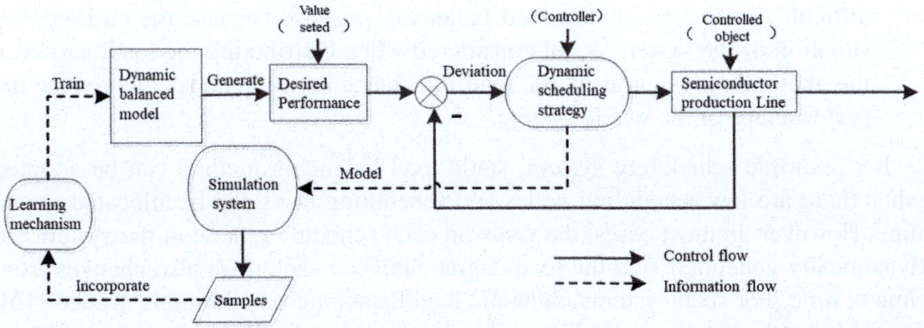

Fig. 7.11 Dynamic scheduling structure of closed-loop optimization of semiconductor production line based on load balancing

(3) The research of dispatching dispatching system (executive mechanism) enables dispatching system to make corresponding dispatching operation according to the instruction of controller.

7.2.2 Load Balancing Technology

Due to the complexity of semiconductor production line processing, the difference in processing capacity of processing equipment and the difference in the number of equipment, there are still many workpieces waiting in line for processing on some equipment (called overload), while other equipment is relatively idle (called light load). In the actual scheduling process, on the one hand, it is necessary to make overloaded equipment complete the workpiece to be processed as quickly as possible and reduce the queue length in front of the equipment; On the other hand, some idle devices can guarantee a certain load and avoid resource waste. Avoid the coexistence of idle and waiting, effectively improve the utilization rate of production line equipment and reduce the average processing time, which is the cause of the load balancing problem [15].

Load balancing is a classic combinatorial optimization problem. Adjusting the load balance of each node of the system is to balance the load of each node by scheduling means, improve the overall equipment utilization rate, and finally improve the system performance. Therefore, adopting an effective scheduling strategy is the key to load balancing.

There are two main types of load balancing adjustment:

(1) Dynamic and static

Static load balancing method usually uses enumeration method, queuing theory and other methods to produce a better scheduling scheme. This method does not refer to the current state of the system and allocates new tasks to each processing node in the system according to the set method. Once the task allocation is completed, the allocation result will not be changed. The mathematical modeling of static load balancing is relatively simple and easy to implement, but it is difficult to achieve a better load balancing purpose because the current load situation of the system is not considered when distributing the load, and even this method may aggravate the load imbalance between nodes and reduce the performance of the whole system.

For a simple scheduling system, static load balancing method can be adopted when there are few scheduling nodes and scheduling tasks can be allocated at one time. However, in most cases, the tasks on each scheduling node in the system are dynamically generated, and the load degree between each node also changes from time to time. For such systems, dynamic load balancing scheduling is needed [16]. Dynamic load balancing scheduling can dynamically distribute the load based on the current load situation of the system, which can improve the utilization rate of the system. However, its implementation is complex, and the scheduling process is computationally intensive, which causes extra overhead to the system to a certain extent.

(2) Local load balancing and global load balancing

Local load balancing refers to only paying attention to the load distribution of some key equipment in the production line, which can meet the load balancing requirements through scheduling, but does not require load balancing for other common equipment, which can reduce the calculation amount and reduce the complexity of the model, especially for the production line including bottleneck equipment [17]. Compared with local balancing, global load balancing emphasizes the establishment of a complete load distribution and scheduling model, considering all nodes as much as possible to pursue the best overall performance, and its implementation is more complex.

Because there are many processing areas in the semiconductor manufacturing process, and the distribution of key processing areas is concentrated, this paper adopts the method of local load balancing, and adjusts the workload distribution in the four processing areas of the production line to achieve the balance requirements.

7.2.3 Selection of Parameters

7.2.3.1 Selection of Processing Zone

This paper takes the 6-inch production line of a semiconductor manufacturer in Shanghai as the background, and its simulation model can track the changes of the actual production line, so as to keep consistent.

The production line of this enterprise includes nine processing areas in total, which are sputtering area, injection area, photoetching area, back thinning area, dry etching area, wet etching area, diffusion area, PVM test area and BMMSTOK microscopic inspection area. Considering that there are many additional work areas in the whole production line and the utilization ratio of each processing area is different in the production process, the local dynamic load balancing method is adopted in the research process, and only dynamic load balancing scheduling is carried out for some processing areas (Table 7.11).

7.2.3.2 Selection of Load Parameters

In the simulation system of semiconductor manufacturing scheduling, a large amount of state information can be obtained during the processing. The attribute sets of the four processing areas are shown in Tables 7.12, 7.13, 7.14 and 7.15.

Table 7.11 Selection of load balancing processing zone

Name	Introduction of processing technology
6-inch lithography area	The most critical step in IC manufacturing is to form the required pattern on the surface of silicon wafer
6-inch dry etching area	The etched surface is anisotropic, with very good sidewall profile control and good etching uniformity
6-inch wet etching area	Has high selectivity to the lower layer materials, does not cause plasma damage to devices, and has simple equipment
6-inch ion implantation area	Accurately control impurity content and penetration degree, so that impurities are evenly distributed

Table 7.12 Lithography area attribute set of 26-inch production line

Serial number	Attribute name	Attribute meaning
17	WIP	Total WIP in lithography area
18	Hotlot%	The ratio of emergency workpieces in lithography area to the total WIP in lithography area
19	Last_1/3_Photo%	Proportion of 1/3 photoetching workpiece number behind photoetching area to total WIP number in photoetching area
20	Bottleneck_M%	Proportion of bottleneck equipment in lithography area to available equipment in lithography area
21	Bottleneck_U%	Bottleneck equipment utilization rate in lithography area
22	Bottleneck_C	Capacity ratio of bottleneck equipment in lithography area
23	RestrainWIP%	Proportion of constrained WIP in total WIP in lithography area
24	Queuing_Job	Queued workpieces in lithography area
25	Queuing_Job_Time	Expected processing time of queued workpieces in lithography area

In order to facilitate the establishment of load balancing model, the selection of state information should be closely related to the scheduling process, and can well reflect the load distribution of the production line. Finally, the following parameters are selected to build the load balancing model of semiconductor production line. The specific parameters are shown in Table 7.16.

7.2.4 Forecasting Model of Load Balancing

Load balance forecasting model is the basis of closed-loop scheduling method based on load balance. In order to realize closed-loop control of production line load, it is necessary to predict the best load distribution in the current state and take this state as a reference state. Only then can the current production line state value be compared

Table 7.13 Attribute set of dry etching zone in 36-inch production line

Serial number	Attribute name	Attribute meaning
44	WIP	Total WIP in dry etching area
45	Hotlot%	Proportion of urgent work pieces in dry etching area to total work-in-process in dry etching area
46	Last_1/3_Photo%	The proportion of 1/3 photoetching workpieces after the dry etching area to the total WIP in the dry etching area
47	Bottleneck_M%	Proportion of bottleneck equipment in dry etching area to available equipment in dry etching area
48	Bottleneck_C	Capacity ratio of bottleneck equipment in dry etching area
49	Bottleneck_U%	Utilization ratio of bottleneck equipment in dry etching area
50	RestrainWIP%	Proportion of constrained WIP in total WIP in dry etching area
51	Queuing_Job	Queued workpieces in dry etching area
52	Queuing_Job_Time	Expected processing time of queued workpieces in dry etching area

Table 7.14 Attribute set of wet etching zone in 46-inch production line

Serial number	Attribute name	Attribute meaning
53	WIP	Total WIP in wet etching area
54	Hotlot%	Proportion of urgent work pieces in wet etching area to total work-in-process in wet etching area
55	Last_1/3_Photo%	Proportion of 1/3 photoetching workpieces after wet etching zone to total WIP in wet etching zone
56	Bottleneck_M%	Proportion of bottleneck equipment in wet etching area to available equipment in wet etching area
57	Bottleneck_C	Capacity ratio of bottleneck equipment in wet etching area
58	Bottleneck_U%	Utilization ratio of bottleneck equipment in wet etching area
59	RestrainWIP%	Proportion of constrained WIP in total WIP in wet etching area
60	Queuing_Job	Queued workpieces in wet etching area
61	Queuing_Job_Time	Expected processing time of queued workpieces in wet etching area

with the reference state to make the production line load distribution tend to the best distributed scheduling and form closed-loop negative feedback.

Table 7.15 Attribute set of injection area of 56-inch production line

Serial number	Attribute name	Attribute meaning
8	WIP	Total WIP quantity in injection area
9	Hotlot%	Proportion of emergency workpieces in injection area to total WIP in injection area
10	Last_1/3_Photo%	The proportion of 1/3 photoetching workpieces after implantation area to the number of products in process in implantation area
11	Bottleneck_M%	Proportion of bottleneck equipment in injection area to available equipment in injection area
12	Bottleneck_C	Capacity ratio of bottleneck equipment in injection area
13	Bottleneck_U%	Utilization ratio of bottleneck equipment in injection area
14	RestrainWIP%	Proportion of constrained WIP in injection area to total WIP
15	Queuing_Job	Queued workpieces in injection area
16	Queuing_Job_Time	Expected processing time of queued workpieces in injection area

Table 7.16 Load balancing parameter attribute set

Serial number	Attribute name	Attribute meaning
1	mov_i	The Mov value of processing zone *i*
2	u_i	Utilization ratio of processing zone *i*
3	hot_i	Number of urgent workpieces in processing area *i*
4	WIP_n	Total number of products of different types of workpieces in the processing area on the production line *i*
5	l_i	Length of queued workpieces in buffer zone of processing zone *i*
6	mov_per_6	MOV in the planned interval of production line (6 h)
7	α	Variable parameters of queuing workpiece information before equipment in DDR scheduling algorithm i
8	α_1	Corresponding to the time required for the workpiece to complete the processing step
9	β_1	Load degree parameter of downstream equipment

In this chapter, the establishment of load balance forecasting model is based on heuristic algorithm, which simulates many production lines, obtains a large number of state information, and selects excellent samples for establishing load balance forecasting model. The model structure is shown in Fig. 7.12.

Among them, the input parts are all 6 h statistical data in the four major processing zones, including the number of moving steps, the average queue length, the average number of emergency workpieces and the equipment utilization rate. A total of 16 parameters are used as the input of the extreme learning machine; The output part is

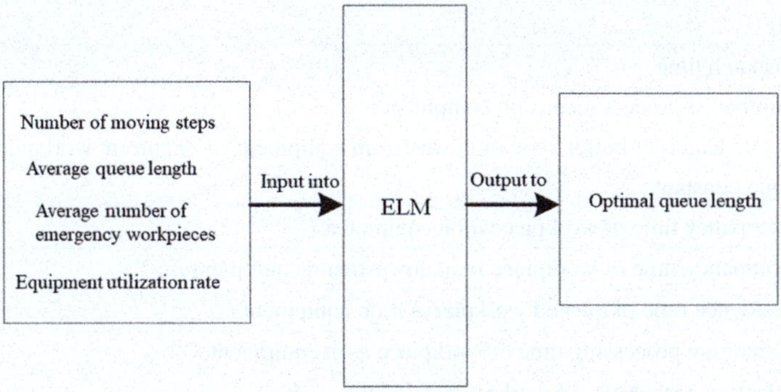

Fig. 7.12 Parameter structure of load balancing prediction model

the optimal queue length of the four processing areas in the current production state of the production line, which will be used in the following scheduling algorithm.

7.2.5 Dynamic Scheduling Algorithm Based on Load Balancing

7.2.5.1 Definition of Algorithm Parameters and Variables

There are many parameters and variables involved in the algorithm, which are summarized as follows:

B_i	Processing capacity of batch processing equipment i
B_{id}	Processing capacity of downstream equipment id
D	There are four processing areas in the production line: lithography area, dry implantation area, wet implantation area and ion implantation area
D^n_{id}	Processing area where downstream equipment id of workpiece n is located
D_n	Delivery time of workpiece n
F_n	The ratio of the actual processing time of the workpiece n is the ratio of the processing cycle to the processing time, in which the processing cycle includes the processing time and the extra time consumption such as waiting in line and transporting the workpiece
i	Available equipment index number
im	Menu index number of equipment i
id	Downstream equipment index number of equipment i
k	Batch index number of queued workpiece group on batch processing equipment i
L^n_{id}	Queue length in the buffer zone of the processing zone to which the downstream equipment id of the workpiece n belongs
$L^{n'}_{id}$	Prediction of the optimal length of queue in the buffer zone of the processing zone to which the downstream equipment id of the workpiece n belongs

(continued)

(continued)

t	Dispatch time
M_i	Number of process menus on equipment i
N_{id}	Queue length of buffer zone of downstream equipment id of current workpiece
O	Time constant
P_i^n	Occupancy time of workpiece n on equipment i
P_{id}^n	Occupancy time of workpiece n on downstream equipment id
Q_i^n	Residence time of queued workpieces n on equipment i
R_i^n	Remaining processing time of workpiece n on equipment i
S_n	Selection probability of workpiece n
T_{id}	Available time of downstream equipment id every day
Γ_k	Selection probability of workpiece group batch k
$\tau_i^n(t)$	The urgency of the equipment i to deal with the workpiece n at any time t
$\tau_{id}^n(t)$	Occupancy of downstream equipment for processing workpieces n at present
x_i^B	Binary variables. If the device i is a bottleneck device at the moment t, $x_i^B = 1$; Otherwise, $x_i^B = 0$
x_{id}^I	Binary variables. If the downstream equipment id is in a light load state at the moment t, $x_{id}^I = 1$; Otherwise, $x_{id}^I = 0$
x_n^H	Binary variables. If the workpiece n is an emergency workpiece at the moment t, $x_n^H = 1$; Otherwise, $x_n^H = 0$
x_n^{im}	Binary variables. If the workpiece n adopts the process menu m on the equipment i, $x_n^{im} = 1$; Otherwise, $x_n^{im} = 0$
$x_{n,im}^{id}$	Binary variables. If the downstream equipment id in the next process of processing the workpiece n is in an idle state at the moment t, and the menu im is used in the equipment i instead of the workpiece, $x_{n,im}^{id} = 1$; Otherwise, $x_{n,im}^{id} = 0$

7.2.5.2 Hypothesis

The following assumptions are made in the implementation process of the scheduling algorithm:

(1) The information required for the scheduling process is completely known, such as the processing time required for workpieces, Work in Process, WIP), etc. this kind of data can be obtained through the manufacturing execution system (MES) or other automation systems of enterprises;
(2) For non-batch processing equipment, in the scheduling decision-making process, we mainly pay attention to the rapid flow of WIP in the production line and the on-time delivery rate of workpieces;

(3) For batch processing equipment, once the batch is completed, the processing time of the batch of workpieces is a fixed value, and the processing time does not change with the number of workpieces;
(4) The scheduling decision of batch processing equipment consists of two steps: Batch processing equipment must first complete the batch processing of workpieces, and the total number of workpieces in each batch must not exceed the maximum value in the process of batch processing. Only workpieces with the same equipment and completely consistent process menus can be processed in the same batch; Calculate the processing priority of each batch of workpieces to be processed. This paper calculates the priority of each batch with the equipment utilization rate in the processing area and the rapid movement of workpieces as the focus;
(5) Once the batch processing type workpieces are completed and enter the processing state, the batch workpieces shall not be changed again before the processing in this step is completed.

7.2.5.3 Decision-Making Process

The decision flow of the scheduling algorithm based on load balancing is shown in Fig. 7.13.

Step 1: Determine the current processing type of equipment. If it is not batch processing equipment, go to step 2; Otherwise, go to Step 6.
Step 2: According to Formula (7.17), calculate the time weight of the workpiece to be processed in the queue of equipment.

$$\tau_i^n(t) = \begin{cases} \text{MAX} & R_i^n \times F_n \geq D_n - t \\ \frac{R_i^n \times F_n}{(D_n - t + 1)} & R_i^n \times F_n < D_n - t \end{cases} \tag{7.17}$$

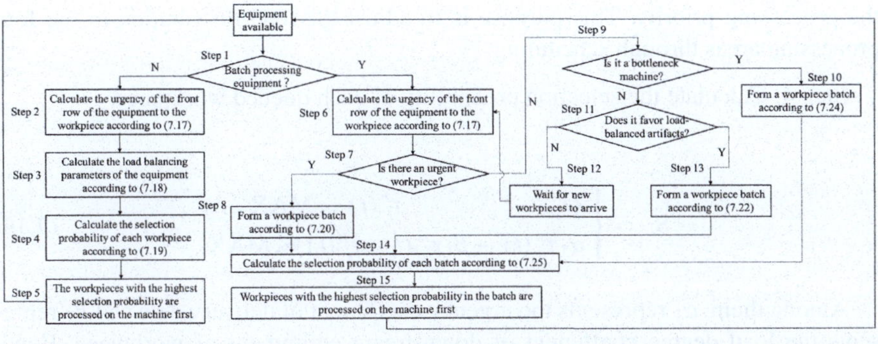

Fig. 7.13 Decision flow based on load balancing scheduling algorithm

Formula (7.17) is based on the delivery time of workpieces, which is used to improve the on-time delivery rate of products. At time t, $R_i^n \times F_n$ refers to the time required for the workpiece to finish processing on the current equipment, which includes pure processing time and waiting time, and $D_n - t$ refers to the actual remaining processing time of the workpiece. If the actual time required for the workpiece to finish processing on the current equipment is not less than the actual remaining time of the workpiece, the workpiece will be set as an emergency workpiece, and it will be processed preferentially in the subsequent processing process, otherwise its time weight will be the ratio of the two, and the greater the value, the more urgent the workpiece will be.

Step 3: Calculate the load balancing parameters of each downstream device.

$$\tau_{id}^n(t) = \begin{cases} \sum P_{id}^n / T_{id} D_{id}^n \notin D \\ \sum P_{id}^n / T_{id} + \left(L_{id}^n / \sum L - L_{id}^{n'} / \sum L' \right) D_{id}^n \in D \end{cases} \tag{7.18}$$

Equation (7.18) indicates the load degree corresponding to the current equipment at time t. The calculation of load degree can be divided into two cases. If $D_{id}^n \notin D$, that is the processing area where the downstream equipment of workpiece n is located does not belong to the four major processing areas D, the load degree of the downstream equipment can be directly calculated; otherwise, a check value needs to be added, which is determined by the ratio of the current queue length of the processing area where the downstream equipment of workpiece n is located to the total queue length of the four major processing areas. It is calculated from the difference between the optimal queue length value of downstream equipment learned by extreme learning machine and the optimal total queue length ratio of the four processing areas, which indicates the difference between the current load distribution ratio and the predicted optimal load ratio. The larger the value, the heavier the load in the processing area where the downstream equipment of the current workpiece n is located, and the lower the load. In the decision-making process, the lower the load in the processing area where the downstream equipment is located, the higher the processing priority. This purpose is to adjust the load distribution in the four processing areas through scheduling.

Step 4: Calculate the selection probability of each queued workpiece.

$$S_n = \begin{cases} Q_i^n & \tau_i^n(t) = \text{MAX} \\ \alpha_1 \tau_i^n(t) - \beta_1 \tau_{id}^n(t) & \tau_i^n(t) \neq \text{MAX} \end{cases} \tag{7.19}$$

Among them, α_1 represents the urgency coefficient of delivery date and β_1 represents the load degree coefficient of downstream equipment of workpiece. Equation (7.19) indicates that at time t, if the current workpiece is an urgent workpiece, its processing sequence will be determined according to the residence time of the workpiece in the queue, which is much larger than the calculated value of the non-urgent workpiece, ensuring the priority processing of the urgent workpiece; If the current workpiece is not urgent, the scheduling priority of the workpiece is determined by the urgency of delivery date, the load of equipment and the load of the processing area where the downstream equipment is located.

Step 5: Dispatch jobs according to the selection probability of the workpieces in the queue calculated in 7. 19, that is, select the workpiece with the highest probability to process on the current equipment.
Step 6: According to Formula (7.17), calculate the time weight of the workpiece to be processed in the queue of equipment.
Step 7: Judging whether there is an urgent workpiece in the waiting queue of the equipment, and if so, turning to step 8; There is no step 9.
Step 8: Batch workpieces according to Formula (7.20).

$$\begin{aligned}&\text{for } \ im = 1 \text{ to } M_i\\&\text{if } \ 0 \le \sum x_n^{im} <\\&B_i \ \text{ then Select}\left\{\min\left\{\left(B_i - \sum x_n^{im}\right), \left(N_{im} - \sum x_n^{im}\right)\right\}\right\}\Big|_{\max\left(Q_i^n\right)}\\&\text{elseif } \ \sum x_n^{im} \ge B_i\\&\text{then Select } \ \{B_i\}|_{\max((R_i^n \times F_n)-(D_n-t))}\end{aligned} \tag{7.20}$$

The purpose of Formula (7.20) is to select common workpieces with the same processing menu as emergency workpieces and group them together. Firstly, count the number of urgent workpieces with process menu im, and judge whether the number of urgent workpieces with process menu im in the current workpieces to be processed is less than the maximum processing capacity of the current batch processing equipment B_i. If it does not exceed, it means that there are few urgent workpieces of this type, and it is necessary to select workpieces with the same menu from ordinary workpieces and group them for processing. If the number of common workpieces in the process menu im is greater than or equal to the number of missing workpieces in this batch $B_i - \sum x_n^H x_n^{im}$, select the required common workpieces to be batched together according to the principle of first-in, first-service, and select all qualified common workpieces to be batched if the number is less than the missing batched quantity; If the number of urgent workpieces in the process menu im has exceeded or equal to the maximum batch B_i, select the urgent workpieces with the most tight delivery time for batch processing. Go to step 14 after the batch is finished.

Step 9: According to the Formula (7.21), judge whether the current equipment is in the bottleneck state. If yes, go to step 10; Otherwise, go to Step 11.

$$\text{If} \sum x_n^{im} \geq (24B_i/\min(P_{im})), \text{then} x_i^B = 1 \tag{7.21}$$

Formula (7.21) indicates that when the queue length before batch processing equipment is greater than or equal to the maximum number of workpieces processed by the equipment in a whole day, the equipment is marked as bottleneck equipment.

Step 10: Batch the machined workpieces according to the Formula (7.22), and then turn to Step 14.

$$\text{Select}\{B_i\}|_{\max(Q_i^n)} \tag{7.22}$$

Formula (7.22) indicates that the workpieces queued in front of the current batch processing equipment are batched according to the process menu. If the number of workpieces meeting the conditions exceeds the single batch processing capacity of the batch processing equipment, multiple batches will be performed according to the principle of first come, first served.i.

Step 11: According to Formula (7.23), determine the load condition of the processing area where the downstream equipment of the workpiece is located. Go to Step 13 if its load parameter value is 1, and go to Step 12 if its load value is 0.

$$\begin{array}{c} \text{if} \quad D_{id}^n \in D \\ \text{if} \quad L_{id}^n/\sum L < L_{id}^{n'}/\sum L', \text{then} x_{id}^I = 1 \\ \text{else} \quad x_{id}^I = 0 \\ \text{elseif} \quad \sum\limits_{im} N_{id} \geq (24B_i/\min(P_{id}^v)), \text{then} x_{id}^I = 1 \end{array} \tag{7.23}$$

Equation (7.23) indicates the load condition of the downstream equipment of the workpiece. If the processing area where the downstream equipment is located belongs to the four major processing areas D, the ratio of queue length in the processing area where the downstream equipment is located to the length of the queue in D is compared with the ratio of optimal queue length in the total predicted length in D. If the actual queue length ratio is less than the predicted queue length ratio, it will indicate that the equipment id is under light load, otherwise it will be heavy load. If the downstream equipment does not belong to D, according to whether the queue length of the downstream equipment of the workpiece exceeds the daily maximum processing capacity of the equipment, if so, the equipment id is in heavy load (bottleneck equipment), and if not, it belongs to light load.

Step 12: Wait for the new workpiece, and then go to Step 8.
Step 13: Batch workpieces according to Formula (7.24).

$$
\begin{aligned}
&\text{for } \; im = 1toM_i \\
&\text{if } \; 0 \le \sum x_{n,im}^{id} < B_i \\
&\text{thenSelect } \; \left\{ \min\left\{ \left(B_i - \sum x_{n,im}^{id} \right), \left(N_{im} - \sum x_{n,im}^{id} \right) \right\} \right\} \Big|_{\max(Q_i^n)} \\
&\text{elseif } \; \sum x_{n,im}^{id} \ge B_i \\
&\text{thenSelect } \; \{B_i\}|_{\max(Q_i^n)}
\end{aligned}
\tag{7.24}
$$

The purpose of Formula (7.24) is to select the workpieces whose equipment required for the next processing step is idle equipment for batch processing. Firstly, count the number of workpieces whose process menu is im and whose next processing equipment is idle equipment, and judge whether the number of workpieces meeting the conditions is less than the maximum processing capacity of the current batch processing equipment B_i. If it does not exceed, it means that there are few urgent workpieces of this type, and it is necessary to select and batch workpieces from the workpieces whose process menu is im and downstream equipment is not light load. If the number of this kind of workpieces is greater than or equal to the number of missing workpieces in this batch $B_i - \sum x_{n,im}^{id}$, select the required workpieces for batch processing according to the principle of first-in-first-service and select all qualified workpieces for batch processing if the number is less than the missing batch processing quantity; If the process menu is im, and the number of workpieces lightly loaded by its downstream equipment has exceeded or equal to the maximum batch B_i, select the workpieces for full batch processing according to the residence time of the workpieces on the equipment. Go to Step14 after the batch is completed.

Step 14: Determine the priority of each batch of workpieces according to Formula (7.25).

$$
\Gamma_k = \alpha_2 \frac{N_{ik}^h}{B_i} + \beta_2 \frac{B_k}{\max(B_k)} - \gamma \frac{\mathrm{P}_i^k}{\max(\mathrm{P}_i^k)} - \sigma \frac{N_{id}^k}{\sum_k N_{id}^k + 1}
\tag{7.25}
$$

where, N_{ik}^h is the number of urgent workpieces in the batch k; B_k is the batch size of the batch k; P_i^k is the occupation time of the batch k on the equipment i; N_{id}^k is the maximum load of downstream equipment for batch processing; Parameters $(\alpha_2, \beta_2, \gamma, \sigma)$ are indicators to measure the relative importance of various types of information.

Formula (7.25) is the formula for calculating the priority of batch workpieces. The first term indicates the proportion of urgent workpieces in the batch, reflecting the on-time delivery rate; The second item represents the ratio of the number of workpieces in the current batch to the number of batches with the largest number of single batches in all batches, reflecting the equipment utilization rate and MOV; The third item represents the ratio of the current batch processing time to the batch processing time with the longest processing time among all batches, reflecting Mov and processing cycle;

The last item indicates the load level of the equipment and reflects the utilization rate of the equipment. In addition, different expected performance indexes can be pursued by adjusting the equal parameter values $(\alpha_2, \beta_2, \gamma, \sigma)$.

Step 15: Select the batch workpieces with the highest selection probability to start processing on the equipment.i

7.2.6 Simulation and Verification

In the load balancing-based semiconductor production line dispatching method, the real-time state information of the production line related to scheduling is encapsulated in the algorithm, including the predicted load ratio of the production line, and then the processing priority of the workpiece is determined by weighting, in which the weighting process involves the weight, that is, $(\alpha_1, \beta_1, \alpha_2, \beta_2, \gamma, \sigma)$, and different performance indexes are obtained by adjusting the weighting parameters.

In this method, we mainly verify the influence of imported load balancing parameters on the performance index of production line, so we use fixed parameters in the verification process. In this verification process, the weighted values are set as follows:$\alpha_1 = 0.5, \beta_1 = 0.5, \alpha_2 = 0.25, \beta_2 = 0.25, \gamma = 0.25, \sigma = 0.25$, and the verification process is divided into two parts, including comparing with traditional heuristic rules under different load conditions of production line. The load conditions are: WIP is light load of 6000 pieces, full load of 7000 pieces and overload of 8000 pieces; And compared with the dynamic dispatching rule without load balancing parameters, the method without load balancing parameters has two main differences compared with the current method:

(1) In the process of load processing of non-batch equipment, the load calculation formulas of equipment in the middle and lower reaches of Step3 are all calculated by Formula (7.26), without considering the relationship between the predicted value and the actual value of load balancing.

$$\tau_{id}^{n}(t) = \sum P_{id}^{n} / T_{id} \tag{7.26}$$

(2) Skip Step11 during batch processing. If bottleneck equipment exists, go to Step10, otherwise go to Step12.

The verification results are shown in Table 7.17. The main performance parameters concerned in this simulation verification process are daily Mov and Equipment Utilization, EU) of the production line. Here, the equipment utilization of the four processing areas is counted in terms of processing areas, which are ion implantation, lithography, dry etching and wet etching. The processing areas contain unused equipment, so the equipment utilization in the whole area is low. In order to ensure the statistical results after the production line is stable, the production data of the first 30 days of the production line has been eliminated, and the production data of the 60 days after the production line is stable are counted in total.

In order to facilitate the comparison of performance indicators, all statistical data are displayed by histogram as shown in Figs. 7.14, 7.15 and 7.16.

The average daily Mov, the average equipment utilization rate (EU) of the four major processing areas and the sunrise film quantity in Fig. 7.14, 7.15 and 7.16 are all treated by normalization method, that is, taking the result of DDRLB method in the figure as the standard, other indexes are compared with this value, and the ratio between each performance index and DDRLB is obtained, so that each performance index can be compared with this method more intuitively. The following conclusions can be drawn:

(1) Under different loads and the same weighted parameters of the algorithm, DDRLB is better than ordinary DDR in various indexes, especially under light load and full load, the average equipment utilization rate of the four processing zones is increased by 2.7% and 2.8% respectively. The closed-loop method is used to adjust the load distribution of the production line to achieve the purpose of improving the equipment utilization rate of the production line. Meanwhile, under light load and full load, the average daily Mov is increased by 1.53% and 1.45% respectively.
(2) When the production line is overloaded, as shown in Fig. 7.16, the performance of DDRLB is not much higher than that of ordinary DDR, because under overload, the equipment of the production line is overloaded and the production capacity of the production line is saturated, so the advantage of load balancing method is not obvious.
(3) In the simulation process, the control variable method is adopted, and DDRLB and DDR methods use the same weighted parameters to compare the changes of production line performance indexes before and after adding load balancing. This parameter can not guarantee that the performance of the current dynamic dispatching method is greatly improved compared with the common heuristic method, and the performance index of DDRLB method is improved compared with the common heuristic method.

7.3 Dynamic Scheduling of Semiconductor Production Line Driven by Performances

To solve the scheduling problem of semiconductor production line in uncertain production environment, a performance index driven Dynamic Dispatching Rule (DDR) based on Extreme Leaning Machine, ELM) is proposed. By means of data mining, the method firstly uses the simulation system of semiconductor production line to simulate, obtains the production data in the production line processing process through simulation, then selects the good parameter group according to the concerned performance index to obtain the required sample set, and then establishes the performance index prediction model of semiconductor production line through the extreme

Table 7.17 Results statistics

Load	Performance/rules	DDRLB	DDR	FIFO	EDD	CR	SPT	LPT	SRPT	LS
Fractional load	Average MOV	29,658	29,204	29,022	29,542	28,931	29,186	28,927	27,444	29,593
	Average EU	0.286	0.278	0.277	0.283	0.275	0.277	0.277	0.255	0.284
	Sunrise film quantity	145	137	142	145	144	136	143	146	143
Full load	Average MOV	29,608	29,179	28,894	29,151	28,541	29,140	28,827	27,350	29,718
	Average EU	0.286	0.278	0.279	0.285	0.270	0.278	0.277	0.254	0.285
	Sunrise film quantity	147	144	138	145	147	139	133	149	142
Overload	Average MOV	29,688	29,495	29,240	29,743	28,696	29,495	29,108	27,453	29,611
	Average EU	0.285	0.281	0.281	0.287	0.271	0.283	0.279	0.253	0.284
	Sunrise film quantity	152	140	129	147	148	139	128	159	146

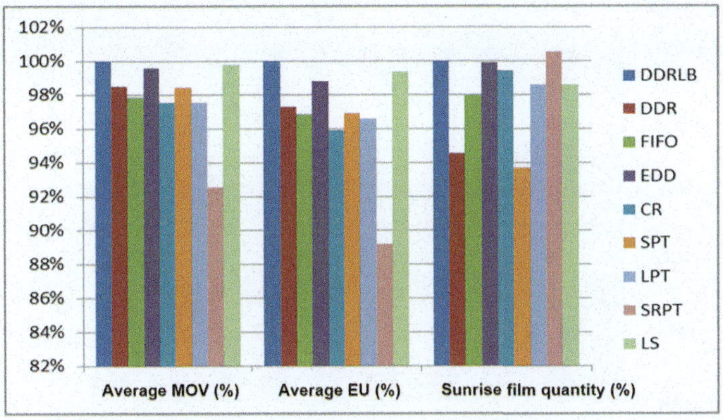

Fig. 7.14 Comparison of scheduling performance under light load condition

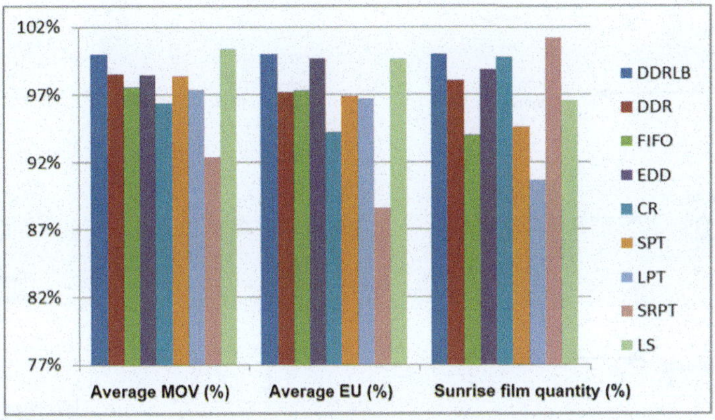

Fig. 7.15 Comparison of scheduling performance under full load condition

learning machine. Then, through the predicted performance index, combined with the real-time state information of the production line, it learns the best parameters needed by DDR algorithm in the scheduling process, and drives the production line dispatching decision, so that its performance index tends to the predicted value. Finally, the method is verified by the simulation platform, which can effectively improve the operation performance of the production line.

7.3.1 Structure of Performance-Driving Scheduling Method

According to the real-time data information of the production line, the performance index-driven scheduling method of semiconductor production line proposed

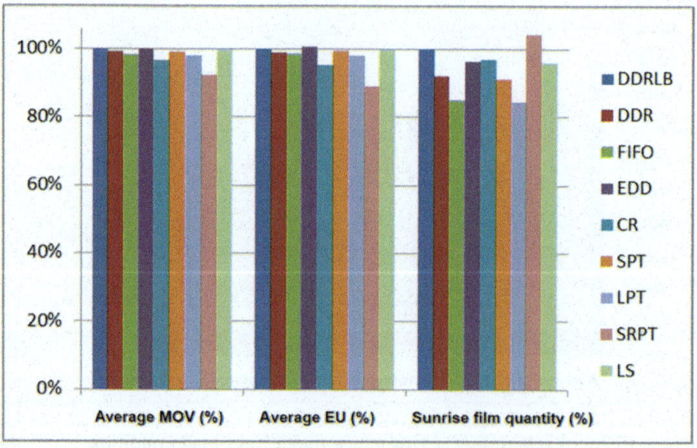

Fig. 7.16 Comparison of scheduling performance under overload condition

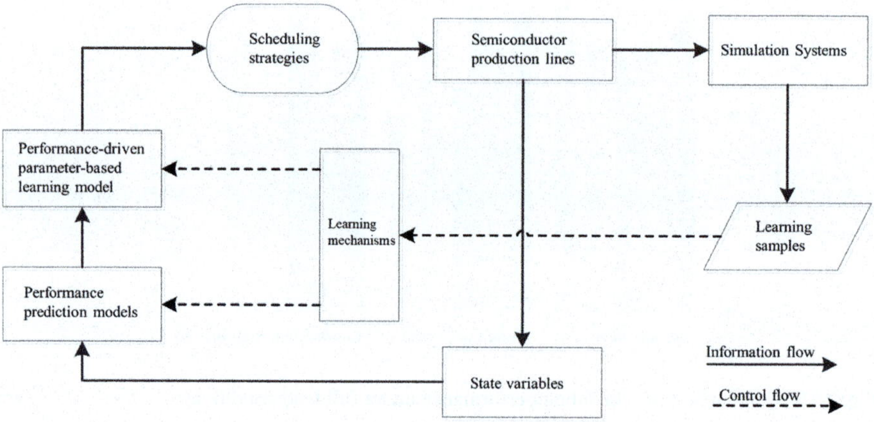

Fig. 7.17 Performance index driven scheduling model

in this paper predicts the best parameters suitable for the production line to make the best dispatch. The model mainly includes the following five parts: simulation system, learning mechanism, performance prediction model, scheduling parameter prediction model and dispatch strategy. The overall structure is shown in Fig. 7.17.

The scheduling model is divided into five parts, including:

(1) Simulation system: The simulation system simulates the production process of semiconductor production line. Detailed real-time status information of production line can be obtained through the simulation model, including queue length of equipment, number of urgent work pieces, queue length of processing zone buffer zone, 6 h Mov value, daily Mov value, etc., and corresponding

performance indicators can be statistically recorded according to different requirements

(2) Learning mechanism: ELM (Extreme Leaning Machine, ELM is adopted as the learning mechanism. Based on many samples generated by the simulation system of semiconductor production line, ELM is used to establish the performance prediction model and parameter learning model of the production line respectively [18]
(3) Performance index prediction model: According to the state information of the production line, take the current state information as the model input, predict the optimal performance index that the production line can achieve in this state, and the predicted parameters will be used as the input of the scheduling parameter prediction model [19]
(4) Scheduling parameter prediction model: According to the predicted optimal performance index, combined with the current state information of the production line as input, the parameters required by Dynamic Dispatching Rule, DDR) in the dispatching decision process are predicted, and the predicted parameters are used in the production line scheduling strategy to guide the production line to dispatch jobs correctly [20]
(5) Scheduling strategy: A dynamic dispatching algorithm based on production line status is used in this paper, which can dynamically make dispatching decisions according to production line status information, and finally make production line performance tend to predict parameter values [21].

In this paper, the performance index prediction model and parameter learning model based on performance index of semiconductor production line are established through Offline learning. In the actual scheduling process, taking 6 h as the time unit, the best performance index that the production line can achieve in the next time unit is predicted through the performance index prediction model based on the production line state information at the last moment of the current time unit, Combined with the status information of the current production line, the parameters needed by DDR method in the process of scheduling decision-making are learned online through the learning model, which can guide the production line to dispatch workers reasonably, and finally promote the production line to achieve the predicted performance index and improve the overall performance of the production line.

7.3.2 Dynamic Dispatching Rules

In order to solve the problem that the traditional heuristic scheduling rules can't consider the real-time situation of the production line, and once the method is fixed, the scheduling can only be performed according to the logic of the method, this paper proposes a dynamic dispatching method DDR (Dynamic Dispatching Rule), which can dynamically schedule the production line by specifying different parameters for the scheduling priority equation according to the different performance indexes concerned in the production line scheduling process.

7.3.2.1 Definition of Algorithm Parameters and Variables

Parameters used in the algorithm are defined as follows:

m_i	The Mov value of processing zone i
u_i	Equipment utilization rate of processing area i
hot	Number of urgent workpieces on the production line
wip_k	The total number of products of different types of workpieces in the four processing zones on the production line
l_i	Length of queued workpieces in buffer zone of processing zone i
Mov_per_6	Next period of production line (6 h)
α	Pre-parameter of queuing workpiece information variables before equipment i in DDR scheduling algorithm
β	Parameters before load degree of downstream equipment

7.3.2.2 Hypothesis

Because this chapter mainly studies the dynamic scheduling of semiconductor production line driven by performance index, the following assumptions are made in solving the dispatching problem:

The following assumptions are made in solving the dispatching problem:

(1) The information related to dispatching is known, such as workpiece processing time, Work-in-Process, WIP) number in front of equipment, equipment available time, etc., and these data can be obtained by MES or other automation systems of enterprises;
(2) The dispatching decision of non-batch processing equipment mainly focuses on the on-time delivery rate of workpieces and the rapid movement of WIP in the production line, which is the focus of this paper. A dynamic dispatching rule DDR is proposed.
(3) For the dispatching decision of batch processing equipment, this paper adopts the common batch method, according to the workpiece processing menu and version number, and after batch processing, FIFO is used for batch processing.

7.3.2.3 Decision-Making Process

DDR is a method to judge the processing priority of the workpiece based on the real-time state of the production line. In this paper, the urgency of the workpiece to be processed and the load degree of the downstream equipment of the workpiece are mainly considered as the judgment criteria of the scheduling priority of the workpiece. In the process of dispatching, the workpiece with the highest priority is selected for priority processing, so as to optimize the overall performance of the production line. The decision-making process is as follows.

Step 1: Calculate the information variables of queued workpieces before the current equipment

$$\tau_i^n(t) = \frac{R_i^n \times F_n}{(D_n - t + 1)} - \frac{P_i^n}{\sum_n P_i^n} \tag{7.27}$$

Formula (7.27) is put forward based on just-in-time delivery rate, in which the P_i^n represents the occupation time of work-in-process on equipment . At time t, the larger the ratio of theoretical remaining processing time to actual remaining processing time of work-in-process in production line, the higher the delay rate, and the priority should be given to its processing in the scheduling process. On the other hand, the occupation time of WIP for the used equipment also affects the information variable value. The shorter the processing time, the higher the information variable value of the workpiece, so that the workpiece can be processed preferentially, which can ensure the rapid flow of WIP in the production line and improve the utilization rate of equipment and the number of moving steps of the workpiece in the production line.

Step 2: Calculate the load degree of the downstream equipment of the workpiece

$$\tau_{id}^n(t) = \frac{\sum P_{id}^n}{T_{id}} \tag{7.28}$$

In Formula (7.28), P_{id}^n represents the occupation time of the workpiece n on its downstream equipment id, and T_{id} represents the theoretical available time of the downstream equipment id every day. At the current time t, the greater the equipment load, the higher the information variable. If the total load of the equipment is greater than all the available processing time in a day when it is in $\tau_i^n(t) \geq 1$, the equipment is considered as bottleneck equipment.

Step 3: Calculate the selection probability of each queued workpiece

$$S_n = \alpha\tau_i^n(t) - \beta\tau_{id}^n(t) \tag{7.29}$$

Among them, the parameter α, β respectively represents the relative importance of the emergency delivery of workpieces and the occupation degree of equipment. Equation (7.29) indicates that at time t, the delivery date of queued workpieces, the occupation degree of equipment and the load of downstream equipment of the workpieces will be considered at the same time in the scheduling process of queued workpieces on the equipment, which will finally enable the workpieces to flow quickly on the production line and improve the overall performance of the production line.

Step 4: According to the calculation result of Formula (7.29), select the workpiece with the highest probability in the queue and process it on the current equipment.

7.3.3 Prediction Model

7.3.3.1 Selection of Model Parameters

Firstly, through the correlation analysis of the processing areas of the production line, the four main processing areas of the production line are selected as the research objects, which are 6-inch injection area, 6-inch lithography area, dry process area and wet process area. In the simulation process, taking 6 h as the planning interval, count the number of moving steps (mov_i), equipment utilization rate (u_i), urgent number of workpieces (hot), the total number of products of different types of workpieces in the four processing areas (wip_k), the number of moving steps in the next planning interval of the production line (Mov_per_6), the priority parameter α, β of workpieces determined in DDR scheduling algorithm and the queue length of real-time buffer workpieces (l_i). As shown in Table 7.18.

Table 7.18 Parameter list of production line prediction model

Serial number	Attribute name	Attribute meaning
1	mov_i	Mov value of in processing zone I
2	u_i	Equipment utilization rate of processing area I
3	hot	Number of urgent workpieces on the production line
4	wip_k	The total number of products of different types of workpieces in the four processing zones on the production line
5	l_i	Workpiece queue leader in buffer zone of processing zone I
6	Mov_per_6	MOV in the planned interval of production line (6 h)
7	α	Variable parameters of queuing workpiece information before equipment I in DDR scheduling algorithm
8	β	Load degree parameter of downstream equipment

7.3.3.2 ELM-Based Prediction Model

Step 1: Sample generation. By adopting DDR method of random assignment, the production line is simulated for 200 days under three working conditions of WIP = 6000, WIP = 7000 and WIP = 8000, and some parameters

(mov_i,u_i,hot,wip_k, MOV_PER_6) required in Table 7.18 are recorded.

Step 2: Sample screening to determine the input and output of ELM. To predict the optimal performance index that can be achieved under the current production line state, the obtained samples are screened according to the selected performance index. This paper mainly focuses on Mov of the production line, so the samples with Mov greater than 8000 within 6 h are selected as the sample set for model building. To ensure the data obtained after the production line is stable, the simulation samples of the first 30 days are removed. Determine the input of ELM, that is, the attribute set of the selected production line under different working conditions; Determine the output of the extreme learning machine, that is, the production line Mov_per_6 in the next planning interval.
Step 3: Determine the parameters of ELM. Determine the number of neurons in the hidden layer of ELM, select the appropriate activation function g(x), and select Sigmoid function here
Step 4: ELM training process. According to the formula $\beta = H^{+}T$, the output weight matrix β is calculated. Because only the output weight matrix β is unknown in the whole training process, the training results of β show that the extreme learning machine model has been trained
Step 5: Select the test set to learn, that is, use extreme learning machine to train the test data and compare it with the test data results.

The main difference between the performance index prediction model and the parameter prediction model lies in the different input and output parameters of the learning machine. In the performance index prediction model, the input parameter is the current attribute value (mov_i, u_i, hot, wip_k) of the production line as the input, and the output value is Mov_per_6 of the production line in the next planning interval, which is used to establish the relationship between the current state information of the production line and the performance index of the next planning interval.
The input parameters in the parameter prediction model include the predicted performance index value (Mov_per_6) predicted by the performance index prediction model and the attribute values (mov_i, u_i, hot, wip_k) of the current production line, and the predicted performance parameters are used in the prediction of production line scheduling parameters.

7.3.4 Simulation and Verification

Taking the 6-inch silicon wafer production line of a semiconductor manufacturing enterprise in Shanghai as the research object, according to the actual needs of the enterprise, combined with the dynamic modeling method, a production line simulation model consistent with the actual production line was built by Siemens Tecnomatix Plant Simulation software as the research platform for simulation verification.

At present, there are nine processing areas in the production line of this enterprise, which are: injection area, lithography area, sputtering area, diffusion area, dry etching area, wet etching area, back thinning area, PVM test area and BMMSTOK microscopic inspection area. The dispatching rules used are based on manual priority scheduling method, referred to as PRIOR for short. The main idea is to set the priority according to the manual experience, to ensure the products can be delivered on time to the greatest extent, that is, to meet the delivery time index.

In the closed-loop dynamic scheduling model adopted in this paper, the dynamic scheduling of the production line is achieved by dynamically generating parameters for DDR method according to the real-time state of the production line. At the same time, the actual Mov of the production line in the current time unit (6 h) is compared with the predicted Mov, and different scheduling algorithms are selected for the production line according to the comparison results, and finally the Mov of the production line is improved. The statistical results are verified in the following three cases: Case1: WIP = 6000, when the production line is light load; Case2: WIP = 7000, when the production line is fully loaded; Case3: WIP = 8000, when the production line is overloaded. In the whole dispatching process, under the conditions of light load, full load and overload, the deviation between actual Mov and predicted Mov in each time unit is less than 10%, which reaches 81.2%, 83.2% and 82.8% respectively.

The comparison between average Mov value and general heuristic rules is shown in Fig. 7.18. In this paper, the Mov results are normalized, and all the data in the statistical results are compared with the maximum value respectively, so that the relationship between each group of data can be displayed more intuitively.

It can be seen from Fig. 7.18 that under three different working conditions: light load, full load and heavy load, DDR algorithm driven by performance index has improved MOV of production line compared with other heuristic rules and compared with the average value of daily average MOV of other heuristic rules, this method has increased by 3.1%, 4.0% and 2.7% respectively.

7.4 Summary

In this chapter, based on two kinds of semiconductor production line models, the corresponding long-term performance index prediction models are put forward, and verified and compared on the same actual semiconductor production line data

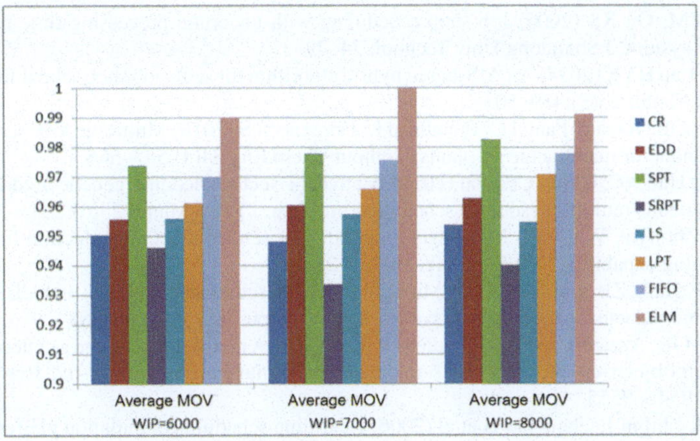

Fig. 7.18 Comparison between DDR algorithm driven by performance index and heuristic rules

set. Aiming at the single bottleneck semiconductor production model, a long-term performance index prediction method based on multiple linear regression method is proposed Aiming at the multi-bottleneck semiconductor production model, a long-term performance index prediction model based on Gaussian process regression method is proposed. On this basis, two different closed-loop scheduling methods, dynamic scheduling of semiconductor production line based on load balancing and dynamic scheduling of semiconductor production line driven by performance index, are put forward, and their effectiveness is verified on the actual production line.

References

1. Xin L, Binghai Z, Zhiqiang L (2009) Event-driven scheduling algorithm for cluster wafer manufacturing equipment. J Shanghai Jiaotong Univ 43(6)
2. Cheng L, Zhibin J, You L, Na L, Na G, Shiqing Y, Wenyou J. Application of rule-based batch equipment scheduling method in semiconductor wafer manufacturing system. J Shanghai Jiaotong Univ 47(2):230–235
3. Guanghui Z, Guohai Z, Wang C, Pingyu J, Yingfeng Z (2009) Dynamic scheduling method of cell manufacturing tasks using real-time production information. J Xi'an Jiaotong Univ 43(11)
4. Tan W, Fan Y, Zhou MC et al (2010) Data-driven service composition in enterprise SOA solutions: a Petri net approach. IEEE Trans Autom Sci Eng 7(3):686–694
5. Hu WJ, Jiuqiang H, Guoji S (2001) Optimization scheduling model of semiconductor manufacturing system. J Syst Simul 13(2):133–135,138
6. Holzinger A, Dehmer M, Jurisica I (2014) Knowledge discovery and interactive data mining in bioinformatics-state-of-the-art, future challenges and research directions. BMC Bioinf 15(6):I1
7. Anzai Y (2012) Pattern recognition and machine learning. Elsevier
8. Blei DM, Ng AY, Jordan MI (2003) Latent dirichlet allocation. J Mach Learn Res 1(3):993–1022
9. Seng JL, Chen TC (2010) An analytic approach to select data mining for business decision. Expert Syst Appl 37(12):8042–8057
10. Li TS, Huang CL, Wu ZY (2006) Data mining using genetic programming for construction of a semiconductor manufacturing yield rate prediction system. J Intell Manuf 17(3):355–361

11. Chen ZM, Gu XS (2005) Job shop scheduling with uncertain processing time based on ant colony system. J Shandong Univ Technol, 74–79
12. Qiu X, Lau HYK (2014) An AIS-based hybrid algorithm for static job shop scheduling problem. J Intell Manuf 25(3):489–503
13. Senties OB, Azzaro-Pantel C, Pibouleau L, Domenech S (2010) Multiobjective scheduling for semiconductor manufacturing plants. Comput Chem Eng 34(4):555–566
14. Wu JZ, Hao XC, Chien CF et al (2012) A novel bi-vector encoding genetic algorithm for the simultaneous multiple resources scheduling problem. J Intell Manuf 23(6):2255–2270
15. Bo Y, Zhongjie W (2007) Multi-agent modeling of semiconductor production line based on machining capability. Syst Simul Technol, 1–28.
16. Lee YF, Jiang ZB, Liu HR (2009) Multiple-objective scheduling and real-time dispatching for the semiconductor manufacturing system. Comput Oper Res 36(3):866–884
17. Senties OB, Azzaro-Pantel C, Pibouleau L et al (2009) A neural network and a genetic algorithm for multiobjective scheduling of semiconductor manufacturing plants. Ind Eng Chem Res 48(21):9546–9555
18. Huai Z, Zhibin J, Chengtao G et al (2006) Real-time scheduling simulation platform of wafer manufacturing system based on EOPN. J Shanghai Jiaotong Univ (Chin Ed) 40(11):1857–1863
19. Amin SH, Zhang G (2013) A multi-objective facility location model for closed-loop supply chain network under uncertain demand and return. Appl Math Model 37(6):4165–4176
20. Yu S, Yiyuan D, Wei J (2009) eM-plant simulation technology course. Beijing: Science Press
21. Haiyan L (2010) Access database management system SQL-server database is different from application. Comput CD Softw Appl 1(5):146–148

Chapter 8
Development Trend of Scheduling Problems for Semiconductor Manufacturing System Under Big-Data

With the advent of the era of big data, how to effectively explore the hidden patterns and rules of the traditional manufacturing industry in the process of acquiring, processing, and analyzing big data to guide and predict the future, so as to realize the value conversion of data, is regarded as the main way to gain competitive advantage in the future. Therefore, the semiconductor manufacturing industry should make full use of its advantages of high automation, informationization and digitalization, and realize the exploration of intelligent manufacturing in the big data environment as a leader. How to effectively acquire, store, analyze and interpret industrial big data, and mine its hidden patterns and rules to guide and predict the future is the key challenge of semiconductor scheduling in big data environment.

8.1 Industry 4.0

Industry 4.0, also known as the fourth industrial revolution, was originally a high-tech strategic initiative of the German government, aiming at ensuring the status of Germany's future industrial production base. Industry 1.0 used mechanical production instead of manual labor for the first time, and the economy and society transformed from agriculture and handicraft industry to a new mode of industrial and mechanical manufacturing driving economic development. However, the mechanical production at this stage was rough and only limited work could be completed. Industry 2.0 develops a new energy power-electricity, which greatly promotes the development of electrical machinery, drives mass production of products, and improves production efficiency. Industry 3.0 is an electronic information age. With the rapid development of information technology based on the Internet, the automation of production is greatly improved, and machines gradually replace human operations. Industry 4.0

L. Li et al., *Data-Driven Scheduling of Semiconductor Manufacturing Systems*, Advanced and Intelligent Manufacturing in China,
https://doi.org/10.1007/978-981-19-7588-2_8

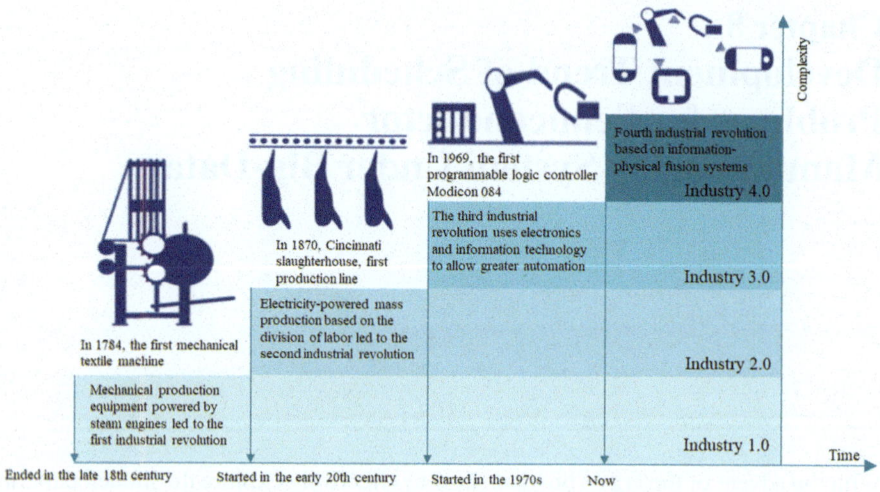

Fig. 8.1 Development course of industrial revolution. (*Source* DFKI 2011)

aims to make full use of the embedded control system to realize the networking and mutual communication of innovative interactive production technologies, that is, the physical information fusion system, and transform the manufacturing industry to intelligence [1], as shown in Fig. 8.1.

The development history of Industry 4.0 is shown in Fig. 8.2. The opportunity of Industry 4.0 is the development of Internet of Things and physical information fusion system. In 1999, Professor Ashton put forward the concept of Internet of Things while studying RFID technology. In 2005, at the World Information Summit, the International Telecommunication Union released ITU Internet Report 2005: Internet of Things. With the help of sensors, embedded technology, network technology and communication technology, the Internet of Things connects, communicates, manages, and controls devices with network interfaces. The Cyber Physical System (CPS), first proposed by the United States in 2006, is defined as a network composed of interactive elements with physical inputs and outputs, which is the product of close coupling between physical devices and the Internet. In July 2010, the German government adopted the High-tech Strategy 2020, which identified industry as one of the top ten future projects [3]. In 2011, at Hannover Messe, Germany, the term Industry 4.0 was first put forward. In October 2012, the Industry 4.0 Working Group, composed of SiegfriedDais of Robert Bosch Co., Ltd., and HenningKagermann of the German Academy of Sciences, put forward the implementation suggestions of Industry 4.0 to the German government. In 2013, the German government formally incorporated Industry 4.0 into the national strategy. Compared with Germany's Industry 4.0, the Chinese government put forward "Made in China 2025" in 2015, aiming at comprehensively improving the manufacturing level and realizing the strategic goal of manufacturing power. The U.S. government put forward the "Industrial Internet" to improve the manufacturing level through digital transformation. A Cyber Physical

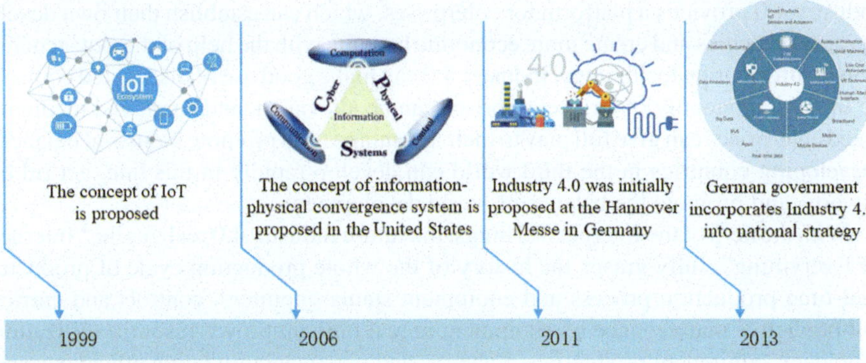

Fig. 8.2 Development history of Industry 4.0

System (CPS) was first proposed by the United States, which is defined as a network composed of interactive elements with physical input and output. It is the product of close coupling between physical devices and the Internet, and the Internet of Things is the embodiment of this system. With the help of sensors, embedded technology, network technology and communication technology, the Internet of Things connects, communicates, and controls devices with network interfaces. With the progress of various technologies, especially the application of 5G technology, the future Internet of Things will realize the "Internet of Everything".

Since Industry 4.0 was put forward, it has made slow progress and faced many challenges in many aspects. Specific challenges come from technical improvements, factory changes, software and hardware platforms and educational level. Industry 4.0 relies on technologies such as industrial Internet of Things, cloud computing, industrial big data, 3D printing, industrial robots, industrial network security, industrial automation, artificial intelligence, etc. However, these technologies are far from mature and applied to industry and are in the initial stage of development. The key communication technology is still in the fourth generation, and the 5G technology is still being researched and developed. The industrial networked control cannot be real-time, and the delayed machine state information also brings difficulties to the control. On the other hand, after decades of development, the organizational structure and working methods of most factories have been fixed. Industry 4.0 requires drastic changes in factories, but the benefits cannot be obtained in the short term, and factories are still in the hesitant stage. At the same time, there is no corresponding mature software and hardware that can monitor the state of industrial products and feed them back to workers to adjust the machines in real time. In addition, Industry 4.0 requires higher education level of factory staff, but there is a gap of interdisciplinary talents in this respect, and universities also lack corresponding professional and skill training.

Challenges also bring many opportunities. Some industries, such as artificial intelligence and service robots, have emerged under the background of Industry 4.0, promoting the development of emerging technologies and mass employment.

Industry 4.0 provides a platform for enterprises, which can establish their own development strategies and create more economic benefits with the help of the new generation of information technology. Industry 4.0 can make good use of distributed control, so that the monopoly of large enterprises can be alleviated, and small and medium-sized enterprises can give full play to their advantages and promote industrial balance. Developing countries in the third world can develop rapidly in this tide, get rid of poverty, and promote the balance of world development.

With the help of the Internet of Things, the future Industry 4.0 realizes the "Internet of Everything", fully grasps the history of the whole production cycle of products, real-time production process and equipment status, monitors, controls and carries out predictive maintenance on equipment in real time, improves resource utilization efficiency, equipment production efficiency and personnel utilization efficiency, and builds a real "smart factory".

8.2 Industrial Big Data

For the industrial field, big data is not a completely unfamiliar term. Since 1980s, the industrial field began to use historical database to manage the data in the production process. With the arrival of Industry 4.0 era, the data generated in the industrial field also shows an explosive growth trend. No matter companies or government agencies, attention to industrial big data is increasing. Although many institutions and scholars have defined big data and industrial big data [4–12], big data is still an abstract concept, and the difference between "big data" and "mass data" is still very vague. Industrial big data refers to massive data that runs through the whole industrial value chain and can realize the rapid development of intelligent manufacturing through technologies such as big data analysis.

The development and application of industrial big data mainly went through the following three stages, as shown in Fig. 8.3:

Fig. 8.3 Development stage of Industrial Big Data

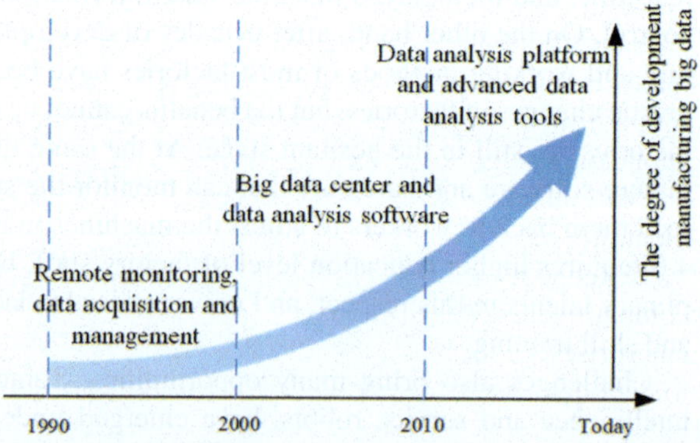

The first stage (1990 ~ 2000): In 1990s, equipment, as an important part of industry, directly affected the economic benefits of enterprises, so once the equipment failed, it would cause huge losses to enterprises. Therefore, the company has developed a product monitoring system based on remote monitoring and data acquisition and management, which monitors products in real time through transmission equipment, greatly reducing the losses caused by failures. OTIS is the largest elevator manufacturing company in the world. In 1998, the company launched REM (Remote Elevator Maintenance), which can not only carry out remote supervision and fault maintenance of elevators, but also contact users in time in case of emergency to ensure the safety of users.

The second stage (2001 ~ 2010): Different from the remote monitoring in the first stage, the second stage adopts the big data center to comprehensively manage the products, mining the value from the data through the data analysis software, and providing the best solution for the use and management of the products. Taking France as an example, influenced by the era of big data, France increased the construction of information systems, and built 16 major data center projects in 2006. Among them, Orange, a subsidiary of France Telecom, uses big data center for data mining and analysis based on highway data detection in France, and provides real-time and accurate road information for vehicles through cloud computing system, thus providing convenience for users to travel.

The third stage (from 2010 to now): the era of "industrial big data". To meet the business needs of industrial big data, the big data center began to transform into a big data analysis platform, which integrates big data integration technology, big data storage technology, big data processing technology, big data analysis technology and big data display technology, and can meet various types of data acquisition and storage, and has the characteristics of high fault tolerance, high security and low cost in performance. At present, the data analysis platform mainly has two forms: tool-based and solution-based. Tool-based platforms, such as LabVIEW-based Watchdog Agent developed by IMS (Intelligence Maintenance System) and NI (National Instruments) of the United States, ensure the correctness of information acquisition with the characteristics of industrial transparency, which is convenient for managers to make correct evaluations; Moreover, it can also meet the requirements of users in different aspects through big data analysis tools and provide them with solutions to problems; GE's tool internet Predix is a typical case of solution-based ecosystem platform. on this platform, developers and users can communicate freely, users put forward their requirements, and developers develop customized data analysis and application solutions according to their needs.

With the increase of information flow in production lines and production equipment, the data volume of manufacturing process and management has soared, and the concept of "driving business development with dynamic data and enhancing core competitiveness of enterprises" has been gradually accepted and valued by most enterprises. In this case, the manufacturing system has changed from energy-driven to data-driven, and data has become a new resource that manufacturing enterprises should pay attention to and make full use of. Therefore, "data-centered" is bound to become an important trend in the further development of manufacturing systems, and the industrial big data analysis method is bound to become the key technology implementation means of intelligent manufacturing.

8.3 Development Trend of Scheduling Problems for Semiconductor Manufacturing Under Big-Data

With the development of information technology, ERP, MES, APC, SCADA, and other information systems have produced abundant data, which contain abundant scheduling related knowledge and can be used to solve complex scheduling problems, that is, using big data technology to extract useful knowledge from related online/offline data, so as to help build a better scheduling model. In fact, the data-based method uses historical knowledge instead of exploring feasible solutions from new data space, which can save a lot of computing resources and computing time.

8.3.1 Data-Based Petri Net

Collect the layout data of production line equipment and product process information from the management system, map them into a time Petri net model, and introduce some heuristic scheduling rules into the model [13]. Mueller et al. [14] proposed a method to transform the data mapping of semiconductor manufacturing system into an object-oriented Petri model. The basic elements of the model include equipment production process, process flow information, equipment, and tool information. This method considers the batch process, downtime of tools and equipment, and rework, which easily leads to oversimplification of the production line, and cannot bring the non-zero state of semiconductor manufacturing system into the model.

8.3.2 Dynamic Simulation

Due to the limitation of simulation software platform, it is difficult to modify the structure of static simulation model to adapt to the physical manufacturing environment. Therefore, based on the static and dynamic information of the production line, the establishment of discrete event simulation model which can reflect the actual processing situation has been widely concerned [15]. The disadvantage of dynamic simulation is that the conversion from data to model is limited in factory simulation, and factory simulation is a special simulation software, that is, the universality of the conversion method needs to be further improved.

8.3.3 *Prediction Model*

Mining the knowledge of various data in manufacturing system through big data technology, and discovering the rules and patterns related to the attributes of production line is helpful to describe the state of production line more accurately and make it consistent with the physical manufacturing environment. Combined with the real-time data generated in the manufacturing process, the future production parameters or performance indicators can be predicted, which is helpful to better guide the production scheduling.

Processing time prediction: Baker et al. [16] recorded the monitoring data of gas flow rate, RF power, temperature, pressure, and DC bias voltage as the input of neural network to predict the running time of ion corrosion process. According to hu et al. [17], using the manufacturing information in MES, a processing time prediction model is established based on support vector mechanism. The results show that the processing time of a work step, including waiting time, equipment adjustment time, pure processing time and visual inspection time, is determined by the state of the machine, the properties of silicon wafers and the operating habits of workers.

Fault occurrence prediction: To predict the occurrence of production line faults and adjust the model configuration, Susto et al. [18] put forward Kalman predictor and particle filter prediction technology using Gaussian kernel density estimation prediction technology and compared their accuracy in monitoring wafer temperature to prevent the production of defective wafers. Kikuta et al. [19] integrated relevant historical data, expert experience, and other information into the knowledge management system, analyzed the average failure recovery time of semiconductor manufacturing equipment, and improved maintenance efficiency.

Prediction of cycle time: Chang et al. [20] combined with self-organizing mapping adopted methods based on case-based reasoning, back propagation network and fuzzy rules, which effectively improved the prediction accuracy of cycle time in semiconductor manufacturing. Meidan et al. [21] used the maximum conditional mutual exclusion method and selective Naive Bayesian classifier to select features and extracted the 20 most important factors affecting the silicon wafer processing cycle from 182 features, which effectively improved the prediction accuracy by nearly 40%.

8.4 Application Example: Big Data Driving Forecasting Model in Complex Manufacturing System

In the wafer manufacturing process, the prediction of workpiece processing cycle is one of the most important tasks for every manufacturer. Accurate prediction of processing cycle can help manufacturers to strengthen their understanding of production line conditions, strengthen the contact with customers, grasp the dynamic market situation and realize sustainable development.

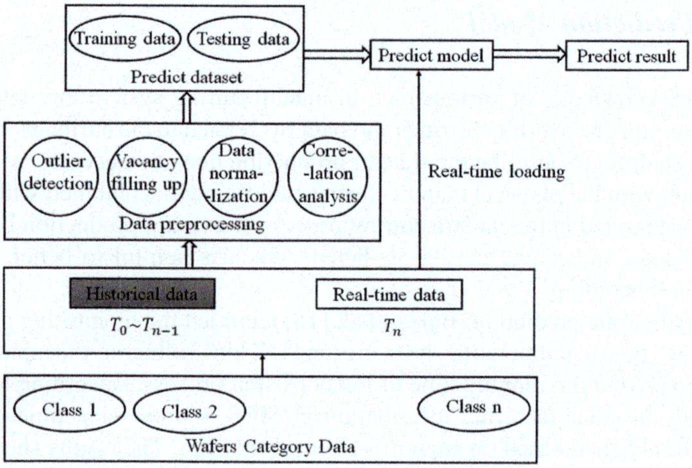

Fig. 8.4 Forecast method framework based on Industrial Big Data

Data management systems with different levels and granularity are deployed in the wafer manufacturing line, which can collect the action data of the lowest-level actuators of the production line, the status information of all resources of the production line in the manufacturing process and manage the business and management information of manufacturing enterprises. The prediction of wafer processing cycle is a practical application of preprocessed industrial big data. The prediction method shown in Fig. 8.4 includes offline training of prediction model and online calling module. When constructing the prediction model of processing cycle based on industrial big data, firstly, the wafers are classified according to the knowledge extracted from the data; Then, the correlation analysis of different types of wafer data is carried out, the input variables corresponding to the processing cycle are selected, and the prediction model is constructed.

Classification and composition of wafer processing cycle: wafer processing cycle refers to the whole time from the time when raw materials are put into production, the scheduled process flow is completed according to the dispatching rules, and the product processing is completed. The factors affecting the processing cycle of wafers are inherent attributes of wafers related to workpieces, wafer processing status, process flow, etc., as well as equipment number, equipment load, WIP, queue length, etc. in processing routes related to production lines.

Wafer processing cycle prediction: the management system of wafer manufacturing transmits the collected data to the production database, in which the initial time of production is T_0, the current time is T_n, the production data from T_0 to T_{n-1} is historical data, and the data from T n is real-time data; After preprocessing the historical data, we can get the data set which can be used for forecasting modeling. After establishing regression relationship with the data in the data set, we can get the corresponding forecasting model. At the same time, we can get the corresponding

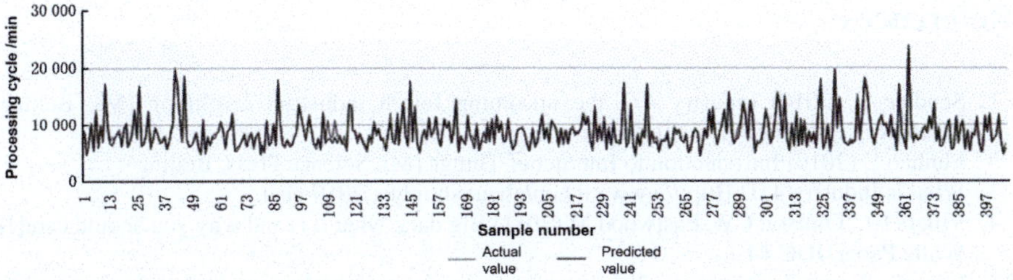

Fig. 8.5 Prediction results of support vector regression

forecasting results by taking the real-time data as input into the regression relationship. Considering the advantages of support vector regression algorithm in dealing with nonlinear and small sample data, the support vector machine is chosen as the regression algorithm.

Taking the production data collected from the mixed production line of 5-inch and 6-inch products of a wafer manufacturing enterprise in Shanghai as the research object, the validity of the above prediction method framework and algorithm is verified. According to the three-month production data of the enterprise, there are hundreds of products with different technological processes in the production line at the same time, and the enterprise data management system has produced 970,286 valid production data. Firstly, according to the inherent properties of wafers, such as wafer size, model, lithography mask version number, technical number, etc. And preprocess that data and sorting out the complete process information of the wafer. Taking the processing cycle of a certain type of wafer as the research sample, there are 45 processes in this type of wafer process, and 406 samples and 224 characteristic variables are generated according to the method in this paper, and 179 characteristic variables remain after dimension reduction; Fig. 8.5 shows the prediction result of processing cycle after establishing the model by 10 times cross validation.

8.5 Summary

This chapter mainly introduces the development trend of semiconductor manufacturing system scheduling in big data environment. This paper starts with the introduction of Industry 4.0, and then introduces the three stages of industrial big data and its development. Based on industry 4.0 and industrial big data, the development trend of semiconductor manufacturing scheduling in big data environment is introduced, including data-based Petri network, dynamic simulation and prediction model; Finally, a concrete application example "Big Data Driven Prediction Model of Complex Manufacturing System" is used to illustrate the semiconductor manufacturing scheduling problem in big data environment.

References

1. Sendler U (2014) Industry 4.0: the upcoming fourth industrial revolution [M]. Beijing: Machinery Industry Press, 2014
2. Yunhao L (2010) Introduction to Internet of Things [M]. Science Press, Beijing
3. What is Industry 4.0? Http://www.gii4.cn/about.shtmlgy, 2019 visit.
4. Villars RL, Olofson CW, Eastwood M (2011) Big data: what it is and why you should care[J]. White Paper, IDC 14
5. Luo S, Wang Z, Wang Z (2013) Big-data analytics: challenges, key technologies and prospects[J]. ZTE Communications 2:11–17
6. Sagiroglu S, Sinanc D (2013) Big data: a review[C]. Collaboration technologies and systems (CTS). In: 2013 international conference on. IEEE, 42–47
7. Wielki J (2013) Implementation of the big data concept in organizations-possibilities, impediments and challenges[C]. Computer Science and Information Systems. IEEE, 85–989.
8. Wan J, Tang S, Li D et al (2017) A manufacturing big data solution for active preventive maintenance[J]. IEEE Trans Industr Inf 2(16):2039–2047
9. Addo-Tenkorang R, Helo PT (2016) Big data applications in operations/supply-chain management: a literature review[J]. Comput Ind Eng 101:528–543
10. Lee J (2015) Industrial big data: the revolutionary transformation and value creation in industry 4.0 era[M]. Beijing: China Machine Press, 5–6
11. Xinjian G, Feng D, Qingmei Y, Zhixiong Y (2015) contents and methods of top-level design of big data in manufacturing industry (part I) [J]. Group Technology and Production Modernization 32(4):12–17
12. Mourtzis D, Vlachou E, Milas N (2016) Industrial big data as a result of IoT adoption in manufacturing[J]. Procedia Cirp 55:290–295
13. Gradišar D, Mušič G (2007) Automated Petri-net modelling based on production management data[J]. Math Comput Model Dyn Syst 13(3):267–290
14. Mueller R, Alexopoulos C, McGinnis LF (2007) Automatic generation of simulation models for semiconductor manufacturing[C]. In: Proceedings of the 39th conference on Winter simulation: 40 years! The best is yet to come. IEEE Press, 2007: 648–657.
15. Ye K, Qiao F, Ma YM (2010) General structure of the semiconductor production scheduling model[C]. Appl Mech Mater Trans Tech Publications 20:465–469
16. Baker MD, Himmel CD, May GS (1995) Time series modeling of reactive ion etching using neural networks[J]. IEEE Trans Semicond Manuf 8(1):62–71
17. Xuechu Z, Fei Q (2014) Processing time prediction method based on SVR in semiconductor manufacturing [j]. J Donghua Univ (English edition) (2):98–101
18. Susto GA, Beghi A, De Luca C (2012) A predictive maintenance system for epitaxy processes based on filtering and prediction techniques[J]. IEEE Trans Semicond Manuf 25(4):638–649
19. Kikuta Y, Tsutahara K, Kinaga T et al (2007) The knowledge management system for the equipment maintenance technology[C]. In: 2007 international symposium on semiconductor manufacturing. IEEE, 2007: 1–4.
20. Chang PC, Liao TW (2006) Combining SOM and fuzzy rule base for flow time prediction in semiconductor manufacturing factory[J]. Appl Soft Comput 6(2):198–206
21. Meidan Y, Lerner B, Rabinowitz G et al (2011) Cycle-time key factor identification and prediction in semiconductor manufacturing using machine learning and data mining[J]. IEEE Trans Semicond Manuf 24(2):237–248

The manufacturer's authorised representative in the EU is Springer Nature Customer Service Centre GmbH, Europaplatz 3, 69115 Heidelberg, Germany. If you have any concerns regarding our products, please contact ProductSafety@springernature.com

Printed and bound by CPI Group (UK) Ltd, Croydon, CR0 4YY
01/10/2026
02225868-0005